ONE PERCENT SAFER

The Secrets to Achieving Safety Excellence
from the World's Finest Thinkers

Compiled, Introduced and Edited
by ***Dr Andrew Sharman***

A NOTE ON PERSPECTIVE

One of the inherent challenges of creating a book with so many contributors is aligning the tone or language to provide the reader with a smooth reading experience. Our position has been to edit very gently, allowing the true voice of each contributor to speak clearly – do keep this in mind as you read. Please note that contributions may not reflect the official views of the organizations to which the contributor belongs – unless this is clearly stated.

For more information please visit **www.onepercentsafer.com**

© Dr Andrew Sharman & Maverick Eagle Press 2020

All rights reserved. No part of this publication may be reproduced, stored in a retrieval system, or transmitted in any form by any means, electronic, mechanical, photocopying, recording or otherwise without the prior permission of the publisher.

Contributors and Andrew Sharman have asserted their moral right under the Copyright, Designs and Patents Act, 1988, to be identified as the authors of this work.

Published by Maverick Eagle Press. **www.maverickeaglepress.com**

British Library Cataloguing in Publication Data. A catalogue record for this book is available from the British Library.

ISBN: 978-0-9929906-4-0

Printed and bound in Great Britain on Forest Stewardship Council © certified paper. This project is supported by Stephens & George Ltd, Merthyr Tydfil, South Wales.

ONE PERCENT SAFER

ACKNOWLEDGEMENTS

This book has been entirely dependent on the goodwill and commitment of many people, not least the contributors who all gave their time, experience and knowledge to share their thoughts in novel yet pragmatic ways.

Extra special thanks are due to James Shannon (www.jshannon.com) who did a masterly job at laying out and designing this book, and to Sarah Thomas for dynamic editing and project management (www.SarahThomasWriter.com).

DEDICATION

This book is dedicated to the giants on whose shoulders I stand, especially Edgar Schein, Charles Handy, Frank Furedi, Geert Hofstede, Jean Piaget and Burrhus Skinner. My gratitude could never be fully explained to you.

This book is also a tribute to those leaders who commit to making the world a little bit safer each and every day.

Let us never forget that we become who we are by what we do.

THE CONTEXT

They told you to follow the rules, to make
new procedures, to sign the policy.
They told you to comply.

They told you it was the law
– enshrined, enacted, enforced.
To do it or there'd be trouble, big trouble:
Fines, courtrooms, jail, even.

They said: "*all accidents are preventable.*"
That they couldn't tolerate any.
They said 'zero' was the only target.
And then there *was* one…

But in your MBA they didn't tell you
how to keep people safe,
nor in your fast-rack executive
development program,
nor in that C-suite retreat.

They never do.
And there's the issue.
They always tell the *what*,
but rarely the *how*, and never the *why*.

THE WHY

So here it is: 2.78 million people died last year due to work accidents or work-related ill-health. No doubt that's an unacceptable number, but it's also hard to get our heads around, so let's break it down.

It's 7,616 people dead every day.
It's 317 every hour.

Count to ten – go on, I dare you.

1, 2, 3, 4, 5, 6, 7, 8, 9, 10

10 seconds, and another person dead.
Another person *just like you*.
A husband, wife, partner, mother, father, brother, sister, son, daughter, friend, colleague.

Another *human*. Just like you.

THE PROBLEM

"Any safety issues?"

We've all heard it. At the end of the Team Talk, the start of the Leadership Team meeting, on the shop-floor Safety Walk… sometimes they even *mean* it, sometimes they actually *want* to know.

But the response is always the same:

NOTHING

Nada. Zilch. Zero.
Silence. A shoulder shrug, a sideways glance, a slight shake of the head.

REALLY???

All these deeply-experienced operators, well-trained managers, high-flying executives – and they have **<u>nothing</u>**…?

And whilst the tumbleweed rolls through, and the clock tick-tocks along, and the manager checks the box to say *'all safe'* another worker dies, somewhere. And then another. Every ten seconds.

And then it's your turn: The phone rings in the middle of the night. The police arrive. Paramedics are already busy. Machines are switched off. People are crying…
The days drag on. The investigators take meticulous notes. The lawyers argue.
You don't sleep…

"Why didn't someone say something? Why did no-one speak up? Why did they stay silent? Why did we wait until someone got hurt?"

THE SECRET

Your ability to follow the rules is
not the secret to your success.
Telling workers to 'be safe' isn't
enough to keep them from harm.

Look, I know that you *care*.

And I know that it's not right that 'the
system' pushes that down inside of you,
and away from the people you care about.

You know that *'good safety = good business'*,
but it's more than that isn't it?
There is a duty not only to do no harm,
but also to make positive change.

It's too bad that so much time has been
lost, but we can't wait any longer.

You have the ability to contribute so much.

We need you, *they* need you, <u>right now.</u>

INTRODUCTION

For too long corporations have stated they want to 'prevent all accidents' and have demanded 'zero injuries' from their people. Sure thing: no-one *wants* to get hurt, but these goals are binary – it's either achieved or not. And as long as we employ human beings, us humans will make mistakes from time to time. So... If you want to ensure your people go home without harm every day, you'll need to maintain a deep distaste for waffle, push back against vagueness, and you'll need to sort the intellectual wheat from the chaff.

As Blaise Pascal said: *"I didn't have time to write a short letter, so I wrote a long one instead."* And there lies the challenge – to simplify thought. As Maurice Saatchi remarked: *"Simplicity is more than a discipline: it is a test. It forces exactitude or it annihilates. It accelerates failure when a cause is weak, and it clarifies and strengthens a cause that is strong."*

But in workplace safety, it's easier to complicate than to simplify. It's easier to elaborate than to distil. To confound rather than to connect. Many of the first contributors to this book baulked at the notion of a 500-word chapter. Several turned in contributions of multiple pages, thousands of words, only to be gently encouraged to try, try and try again[1]. The International Labour Organization reckon that the annual global cost of work-related injuries and deaths totals around $3 trillion – that's at least 4% of the world's Gross Domestic Product (GDP). The European Agency for Safety & Health at Work say that it's at least EUR 476 billion every year in Europe alone. Over in America, the Occupational Safety & Health Administration calculate that the 23,000 on-the-job injuries that occur each day add up to $250 billion annually. And the World Monetary Fund reports that this annual cost to US business of workplace injuries is greater that the GDP of 91 countries.

The *Centers for Disease Control & Prevention* ('CDC') suggest that a fatal injury carries an average cost of around a million dollars, whilst the *National Safety Council* (of the USA) reckons that adding in the indirect costs (such as workplace disruption, loss of productivity, worker replacement, increased insurance premiums, and legal fees) tips the figure much further north, to around $3 million.

The organization for which I serve as President, the *Institution of Occupational Safety & Health*, reports that benefits of a safer workplace include a happier and healthier workforce, lower staff turnover, improved productivity and a better corporate reputation. But there's more than that: the *National Safety Council* and *Centers for Disease Control* believe that for every dollar spent on improving workplace safety, the Return-On-Investment is between four and six times. So good safety *really is* good business.

Whoa!!!! Hang on: it's not about numbers. It's about people, *real* **people.**

We can't fix everything all at once. But we can do *something* right now. Just do *something*[2].

Just do one thing.

Just make your organization **one percent safer**. If we all do just one thing, just improve things by one percent, then that's 28,000 people that get to go home without harm, instead of ending up dead – *each year*. 28,000 husbands, wives, partners, mothers, fathers, brothers, sisters, sons, daughters, friends, and co-workers. **28,000 humans.**

This book is your toolkit: 100^3 of the world's most thoughtful leaders, right here just for you. Read them all, read just one, skip some, read what you prefer.

Just do something.
One thing. One idea.

ONE PERCENT SAFER.

---

1. Acknowledging my heritage, this is a nod to Robert the Bruce, who, just before the Battle of Bannockburn in 1314 roused his troops and told them: *"Be strong and bold, without fear, and you will surely have victory."* It's a pretty good call-to-arms today, too, isn't it?
2. Okay, it's not quite as catchy, but it's a start!
3. Actually, that *was* the plan, and then others wanted in. So if you count 'em, you'll find there's actually 137 contributions in the book now!

CONTRIBUTOR LIST

Sameh Abadir
Jason Anker
Stian Antonsen
Andrew Barrett
Francois Barton
Robert Bea
Eduardo Blanco-Munoz
Simon Bliss
Christian Boehmer
Joachim Breuer
Tony Brock
Martin Bromily
Urbain Bruyere
René Carayol
Tristan Casey
Edmund Cheong Peck Huang
Lester Claravall
Theo Compernolle
Todd Conklin
Sam Conniff
Cary Cooper
Dominic Cooper
Mike Cosman
Andi Csontos
Anne Davies
Shaun Davis
Kimberley de Selincourt
Sidney Dekker
Philippe Delquié
Ruth Denyer
Bernie Doyle
Ockert Dupper
Walter Eichendorf
Tim Eldridge
Jake Esman
Grigorii Fainburg

Kristin Ferguson
Rhona Flin
Gerald Forlin
Ralf Franke
Frank Furedi
Ron Gantt
E. Scott Geller
Francois Germain
Isaac Getz
Alistair Gibb
Gerd Gigerenzer
Christophe Gillet
Fred Goede
David Gold
Marshall Goldsmith
Sonni Gopal
Jop Groeneweg
Gudela Grote
Frank Guldenmund
Judith Hackitt
Andrew Hale
Charles Hampden-Turner
Ian Hart
Jimmy Haslam
Leandro Herrero
Vincent Ho
David Hofmann
Gert Jan Hofstede
Erik Hollnagel
Andrew Hopkins
Patrick Hudson
Jon Hughes
Stuart Hughes
Rod Hunt
Ehi Iden
Mieke Jacobs

Phil James
Nektarios Karanikas
Trish Kerin
Evelyn Kortum
Phil La Duke
Jean-Christophe Le Coze
Bridget Leathley
David Liddle
Peggy Linder
Maria Lindholm
Natalie Lotzmann
Jim Loud
Joan Lurie
David Magee
Erik Matton
Karen McDonnell
Kathryn Mearns
Thierry Meyer
David Michaels
Luiz Montenegro
Alex Morales
Keith Morton
Sanjay Munnoo
Kevin Myers
Marcin Nazaruk
Michael O'Toole
Diane Parker
Phillip Pearson
Sally Percy
Gerrit Poggenpohl
James Pomoroy
Drew Rae
Anthony Renshaw
Rob Richardson
Norman Ritchie
Ivan Robertson

Fredrik Rosengren
Deborah Rowland
David Sarkus
Edgar Schein
Peter Schein
Davide Scotti
Kathy A. Seabrook
Andrew Sharman
Steven Shorrock
Mark Simmonds
Karl Simons
Paul Slovic
Carole Spiers
Christiane Spitzmueller
Malcolm Staves
Rob Stephenson
Neal Stone
Edwin Stoop
Brian Sutton
Darren Sutton
Jukka Takala
Anna-Maria Teperi
Sven Timm
Bernd Treichel
Nick Turner
Davide Vassallo
Bruno Vercken
Louise Ward
Lawrence Waterman
John Mark Williams
Ken Woodward
Louis Wustemann
Paul Zonneveld
Gerard Zwetsloot

A DISCLAIMER AND SOME ASSUMPTIONS

Look, this book might not work for you. It may not hit the mark – or it may hit it too hard. It's certainly taken me outside my comfort zone in creating it. People said I was crazy with this idea, and maybe they were right. Assembling everyone whose opinions I value, in one book, asking them to distil years of wisdom to just one page each, yep, it *is* nuts. But it's done now, so come on, get out of your comfort zone too, come with me on this journey to making the world of work a better place to be human.

This book isn't just a nice read: it isn't *Red Riding Hood, Mark Twain* or *John le Carre*. Every contributor to this book wants to help you see something you may have missed – beyond the rules and regulations, looking past the audits and checklists, more than just asking *"Got any safety issues?"*

This book pivots on two simple assumptions on my part:

1. you know how to be human
2. you don't want your folks to get hurt or die at work

I think most folks will agree with at least the second one of these, but you never know. Those who disagree need read no further.

Allied to that second assumption, I believe that if it's in our power to prevent something very bad from happening (such as an injury, fatality, or serious ill-health) without sacrificing anything of comparable moral importance, we ought, morally, to do it.

Oh, hang on, there's maybe another:

3. you don't need to be taught how to be a leader, you might just need a little bit of inspiration and the permission to *lead with safety*

You got it! *Go on, then.*

Go on, then…

...CHANGES
EVERYTHING

THE PERFECT STORM

SAMEH ABADIR

Common sense leadership in crisis is often not a common practice.

Abraham Lincoln once said, "Nearly all men can stand adversity, but if you want to test a man's character, give him power." Author George Orwell stated something along the same lines: "The real test of character is how well you treat someone that has no possibility of doing you any good." Both quotes say a lot about how others view our actions and decisions during times of crisis.

Presidents typically say they want to be surrounded by strong-willed people who have the courage to disagree with them. As President-elect, Barack Obama, who reached out to former rival Hillary Rodham Clinton and Republicans across the aisle, might have actually meant it. Abraham Lincoln meant it. He appointed his bitter adversaries to crucial posts, choosing as war secretary a man who had called him a "long-armed ape who does not know anything and can do you no good."

Today's corporate leaders continue to be challenged with how to manage external and internal risks that present in complex and interconnected ways. The corporate world has seen its fair share of crises, and any risk can quickly turn into a crisis. By paying attention to a few dos and don'ts, you may be able to preserve your company's value and its reputation.

History has shown us that corporations have taken various approaches in how they responded to crises. Some responses have been more successful than others. Regardless of how boards choose to respond to crises, their decisions could have a significant bearing on the company's stock and its reputation. It's vital that corporate leaders think through the potential ramifications of their decisions carefully.

Right now, I think we are seeing a perfect storm. Different forces are aligning here:

One is scarcity, the other is loss of control or feeling lonely. When we combine all of these forces, we may be falling into the hidden trap of crisis: losing our fighting spirit culture after a long confinement period. Soldiers living too long in trenches or managers spending too long hours in the office both lose their appetite, but they also their ground engagement capabilities.

Experienced leaders will have to prepare for crisis by building and maintaining their organizations' social capital at times of strength and stability, so when crises happen, no one feels alone, losing control, excluded or helpless in his or her trenches.

Are you, in your organization, building social capital and reaching out to those you *don't* need <u>before</u> the storm hits?

Dr Sameh Abadir is Professor of Leadership & Negotiation at IMD Business School and the Co-Director of IMD's *Negotiating for Value Creation* program. His field of expertise is organizational behaviour. Abadir is the founder of the Leadership and Safety Culture program and the Master in Business Excellence at CEDEP.

(DON'T) BE LIKE ME

JASON ANKER

In 1993 I was involved in a totally avoidable accident on a construction site leaving me paralyzed from the waist down after falling just 10 feet (3m) from an unsecured ladder. At 24 years old this had a huge impact on me, not just physically but also mentally. My marriage quickly broke up and my wife walked out taking my two children.

Unable to speak about my feelings, I bottled them up mainly with the use of alcohol and anti-depressants In 1995, following an unintentional excessive dose of 'recreational drugs' I spent 17 days in a coma fighting for my life.

This should have been the turning point in my life; however, I spent the next 14 years 'just about coping', struggling with day-to-day life and raising my two children who were now living with me. Then in 2008, a chance meeting literally changed the direction of my life. This chap had heard about my story and he encouraged me to take my story out to industry and not just share my accident but more on the impact post-accident, on my life, and also my family and friends. The rest is history and since then I have shared it over 2,500 times in the UK, Europe and Worldwide with an inspirational/motivational *'please do not be like me'* type of presentation.

In the last few years I've developed a new approach to my talks, a more *'be like me'* presentation: giving focus to the positive perspectives on my life, especially as there is now – and rightly so – a more holistic approach to Health, Safety and Wellbeing. My new talk, along with writing a book, made me revisit my accident and understand the key factors: my not speaking up, the unplanned work, etc – all the usual suspects were involved, but it also went deeper into the culture of the organization and also my mindset at that time.

I now talk about my life a year prior to my accident, how my life had started to fall apart, the recession of the early 90's, being made redundant and then working in jobs I didn't want to do just to earn money. This had, not knowing at the time, a huge impact on my wellbeing, and I can now see the spiral of events that led up to my accident had started way before January 3rd 1993.

I draw many similarities between the situation I was in prior to my accident and where we are today, as we start to emerge from the Covid19 pandemic: job insecurity, pressure to meet deadlines, etc. So as 'leaders', what is it that we can do to stop the anticipated spike in accidents.

From my viewpoint as an accident victim it starts with something low-cost, simple, yet highly effective. Whatever your position in the organization, you are never too important to be nice to people. Ask workers how they're doing, sincerely, and take an interest in their reply. Whilst the notion of 'social distancing' has meant to keep us apart physically, it has actually given us the chance to start communicating more. For me, that's a great thing.

Jason Anker has been a motivational safety speaker since 2009. After a catastrophic accident in 1993 leaving him paralyzed, he has spoken around the world on health, safety, wellbeing and resilience. Anker was awarded an MBE in 2015 for services to the Construction Industry and in 2018, he published his first book.

Be like me

THE NOWHERE MEN OF (SAFETY) MANAGEMENT

STIAN ANTONSEN

He's a real nowhere man
Sitting in his nowhere land
Making all his nowhere plans for nobody

John Lennon never meant to write a song about safety management. Yet, to those interested in the management of safety, the lyrics of The Beatles' 'Nowhere Man' provides a metaphor for what I believe is a fundamental problem with management in general, and safety management in particular: The growing distance between management – focusing on planning processes and governing principles – and the people at the sharp end of organizations.

Somewhere in the development of management studies and safety management principles, the preoccupation with organizational structures, large-scale organizational change, management systems and general risk assessment processes, the actual high-hazard work faded into the background. Parallel to this, or perhaps because of this, management became a discipline of its own. The mantra seemed to be that a good manager can be set to manage just about everything, irrespective of the characteristics of the organization's core activities and the potential risks involved.

Nowhere man please listen
You don't know what you're missing
Nowhere man, the world is at your command

Why is this a problem for safety management? It is a problem because knowledge and competence influences what you are able to see, understand, put into words and pay attention to. Without having a real understanding of the work that goes on, the technology involved and the constant conflicting objectives of operational work, how can managers contribute to detecting small problems and anomalies before they turn into dangerous situations? It is a problem, because managers have power to make decisions, and without understanding operations and technology they risk making really bad ones.

First and foremost, it is a problem because it creates an organizational gap between management and operations. This is when managers become "nowhere men" far from the operational realities of organizations, and it becomes hard to maintain the sensitivity to how operations are being performed and what can go wrong while performing them. This gap can be a form of organizational blindness, where the sources of success and failure becomes increasingly invisible to those with power to change the organization. This, in turn, can result in decisions eroding old foundations for success and creating new conditions for failure.

He's as blind as he can be
Just sees what he wants to see
Nowhere man, can you see me at all?

The argument here is both basic and old. Virtually every accident investigation and every theory of organizational safety in one way or another emphasizes the importance of the connections between the management sphere and the operational sphere of organizations. Still, there is a powerful drift toward making managers "nowhere men," preoccupied with the abstract processes of *organization*, leaving less room for maintaining the sensitivity to *operation*. The ability to remain sensitive to operations is not an abstract organizational quality – it is a competence and a practice. This competence needs to be acknowledged, made visible and nurtured as a basic building block of (safety) management.

Stian Antonsen is research professor at NTNU Social Research, and adjunct professor in safety management at the Norwegian University of Science and Technology. He is the author of several books and articles on topics related to safety culture and management.

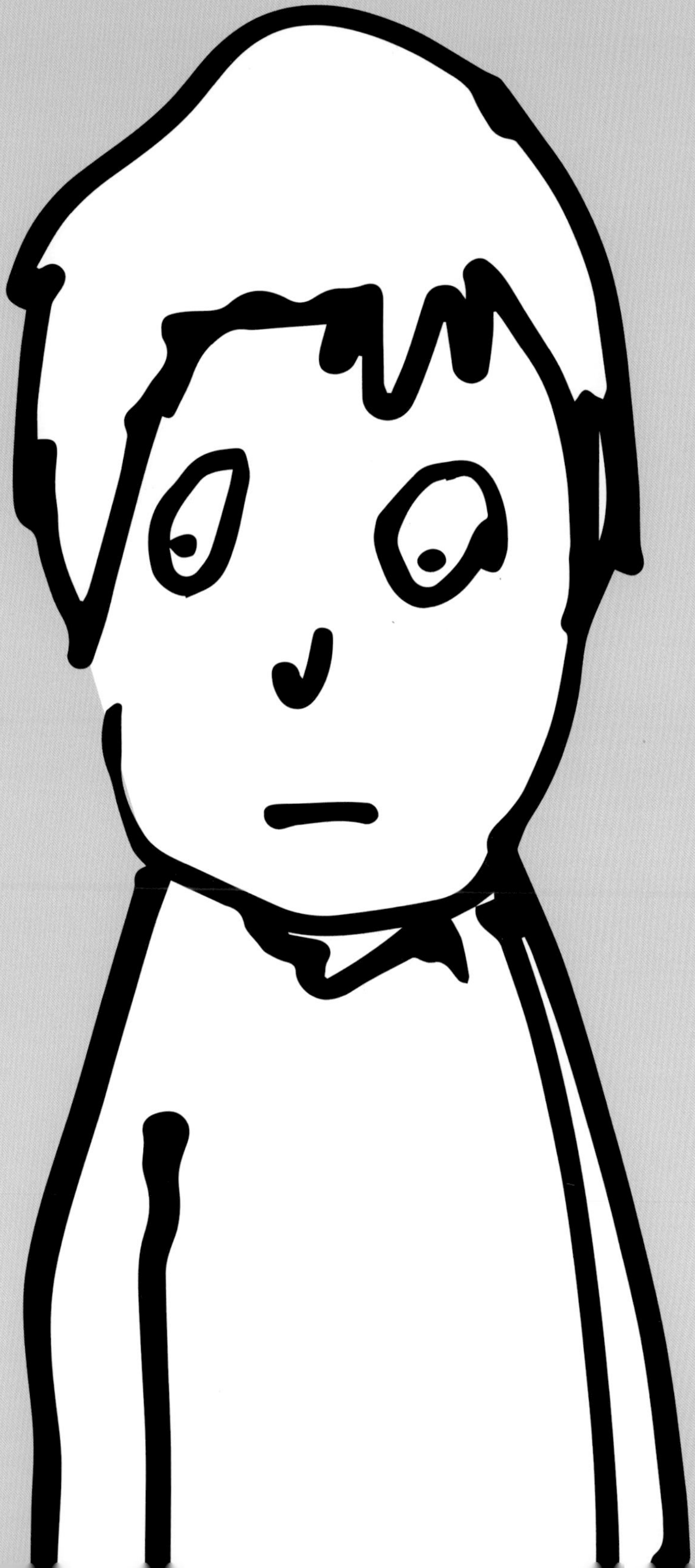

DON'T CHASE RATS AND MICE

FOLLOW THE ELEPHANTS

ANDREW BARRETT

Your job as a leader is to enable your people to be the best at what they do. Don't expect your people to succeed and be safe with sub-optimal or broken equipment, systems or resources. Real leaders fix things, provide the resources and solve problems when workers can't themselves.

As a leader, be obsessive about the most serious risks, and what you and your people need to do to anticipate and respond to these. Entrust all the other risks to your people to manage, and they will do this capably.

In other words: Follow the elephants intently; don't chase rats and mice.

An investment in a safety control (like a lifting device or machine guard) pays you back. The first payback is in reduced risk. But the far bigger payback is the commitment and care your people will reciprocate when they know you are committed to and care for them. Think about this as the dividend.

Businesses are no longer like production lines in a factory—even factories aren't like that anymore. Work is not linear; it is messy and complex. Workers are not interchangeable parts; they are adaptive and creative human beings. Your approach to your business, your people and their safety needs to reflect the current reality, not as it was in 1900.

In other words, no-one reads lengthy documents. An "incident investigation" sounds like a police activity (which no one wants to be part of), yet a "learning conversation" might achieve the same aim, but give you a better outcome.

People spend a lot of time at work. What if you created a workplace where workers loved to be? Wouldn't that be safer, happier, healthier? That is perhaps the number one priority of a leader.

Compliance is a race to someone else's minimum standard. Health and safety are ongoing outcomes of a successful, people-oriented business. Focus on the latter, and the former largely sorts itself out.

How do you respond to bad news? Care for the hurt and welcome the news: you now have something to learn from. What if things rarely go wrong?

I doubt that's actually the case—it's more likely you just don't find out.

How can you effectively lead when you don't know what's really going on?

Fortunately, things go well most of the time. Shouldn't you proportionately pay more attention to learning from how things go well? In other words, "normal" is almost universally ignored. It's worth paying more attention to.

Blame and punishment fix nothing. Blame and punishment create fear; they squash learning; they won't actually change anything or make you feel better anyway.

An organization which is not learning, and acting on this learning, can never thrive, and is unlikely to survive. Assuming the people who do the work are the experts—which they are—your job as a leader is to facilitate that learning within and across your organization. Call it continuous improvement. Call it strengths-based leadership. Call it whatever. I call it smart.

Andrew Barrett coaches business leaders to get better business outcomes by becoming better people leaders. Health and safety is a big part of this, but making work better overall makes for better business performance in lots of different ways.

IF IT FEELS
GOOD, DO IT

MAGPIE OR FUNDAMENTALIST?

FRANCOIS BARTON

"I say 'if it feels good, do it'."
– Police Chief Wiggum, 'The Simpsons'

I came to health and safety by chance, not calling. I joined the New Zealand health and safety regulator to understand how "enforceable duties" could drive behaviour. From that initial focus grew an interest in "safety culture;" I have since been exposed to numerous approaches, and my "pallet of safety frameworks" has grown.

Over that time I've worked with many senior leaders who had to confront appalling safety performance, and managed to turn the business around to the point where it was productive and its people were thriving.

I didn't see one single tool or approach deliver those improvements. Indeed, almost all of these approaches delivered some value in some contexts. What was common, however, were two 'model-agnostic' components:

- *Conscious curiosity by senior leaders*
  A core recognition that there are answers and experiences beyond the reach of the leadership team that need to be invited and enlisted from the across the business.
- *Core belief in people's dignity*
  Regardless of context, people want to be safe, acknowledged and understood – this is human dignity.

When I see organizations succeed – it has universally been anchored in these two ingredients.

The challenge for senior leaders, though, is when they seek to make sense of their role in safe, healthy and productive work, they are confronted with an intellectual "marketplace" of proprietary or mutually exclusive models. They are confronted with what are promoted as "correct answers" that are often in conflict with each other.

I have seen the impact that this "marketplace" can have on organizations wanting to improve. We end up with almost religious debates about the merits and sins of differing approaches, zealous adoption of one approach and equally vigorous rejection of others, or a confused interpretation of what some approaches actually mean.

Yet I have heard a well-intended, committed and capable CEO say to me that they were ready to "mature *beyond* Safety I to Safety II so we can celebrate success." One does not demand the rejection of the other; this leads to the value of both approaches being lost.

The rich and ever-evolving range of safety approaches gives us smart tools and methods to deliver improvement. But those approaches need to be understood and deployed in a complex, diverse and dynamic world. That's why we must apply these tools with conscious curiosity of our context and treat the people we lead and work alongside with human dignity.

So what's my practical idea to help senior leaders play their part in contributing to our collective 1% improvement? Be a magpie, not a fundamentalist – a curious and dignified magpie. There is no single "right way" for to you lead and enlist your people in driving better work where people can thrive. Just start, stay curious and listen to your people.

As Chief Wiggum so wisely observed – *if it feels good, do it.*

Francois Barton is the Executive Director of the Business Leaders' Health and Safety Forum, New Zealand.

THE FIVE Cs

ROBERT BEA

Of particular importance near the end of my career was a 10-year duration series of research and development projects that addressed implementation of advanced Risk Assessment and Management (RAM) processes to substantially reduce the risks and improve the safety of infrastructure systems that existed in high hazard environments.

A key result from these projects was the identification of the five organizational components ('5 Cs') required for successful implementation of advanced system RAM processes:

1. Cognizance – The responsible organizations (commercial-industrial AND public governmental) must develop a realistic awareness of the hazards and risks that confront their infrastructure systems. Concern about safety and risks is constant. Awareness is crucial. Diligence to maintain systems with 'As Low As Reasonably Practicable' (ALARP) risks during their entire life-cycle is even more critical.
2. Commitment – The management and operating personnel must develop a sustained 'top down and bottom up' commitment from those involved, so the necessary resources (human, organizational, monetary, knowledge, experience, physical, environmental) are provided to enable effective application of ALARP RAM 'Barriers' (integrated proactive, reactive, and interactive processes) to enable development and maintenance of systems that have ALARP risks.
3. Culture – The beliefs, values, feelings and resource allocation and utilization processes of the organizations must be one devoted to "knowing what is right, and doing it right," consistently delivering systems that have ALARP risks and understanding that these efforts require constant vigilance, diligence and continuous improvements. This 'C' can be identified as an effective comprehensive RAM based 'Safety Culture.'
4. Capabilities – The human, organizational, and other parts of the infrastructure systems – including combinations of human operators, responsible organizations, hardware, structures, environments, standards and guidelines, and the interfaces between these interconnected components – must be highly developed, integrated and 'excellent' (continually improving). Thus, the proven principles of RAM technology can be properly and effectively developed and implemented. These efforts focus on continuous improvements to enable realization of the different kinds of benefits from application of RAM technology.
5. Counting – This is a very important 'C.' Counting means development of validated quantified metrics that can be used by regulators, managers, engineers, operations and maintenance personnel to help evaluate system risks (likelihoods and consequences of major failures) throughout the life-cycle of a system. These risk assessments do not only address 'natural hazards' (e.g. storms, droughts, pandemics, earthquakes, etc.), they also address the very important human and organizational factors associated with task performance, decision making, and knowledge development and utilization. Risks that are ALARP are based on quantitative monetary cost-benefit evaluations that include proper recognition of both short-term and long-term monetary costs, standards-of-practice evaluations, historic precedents and national and international standards and guidelines for determination of ALARP risks. What is effectively measured can be more effectively managed.

These projects demonstrated that if one or more of the '5 Cs' were defective, then failure of the Risk Assessment and Management (RAM) processes could be expected.

Dr Robert Bea, Emeritus Professor at University of California Berkeley's Center for Catastrophic Risk Management, has devoted his 66-year career to practice, research, teaching and service associated with infrastructure system Risk Assessment and Management (RAM) in 26 countries with engineering, construction, operations, maintenance and management disciplines associated with complex infrastructure systems.

5

WHY PEOPLE TAKE CHANCES

EDUARDO BLANCO-MUNOZ

Do you remember the classic scene from the cartoons of our childhood, where the hesitating character is torn between two choices, represented by an angel on one shoulder and a devil on the other?

After the accident, it's very easy to blame the poor soul who took that 'obviously stupid' shortcut. It's almost as easy as having discounted beforehand that nobody in their right mind would do such a thing…

But you know what? People *do* take risks – look first at yourself, then up, down and around. Everybody, every day takes risks. As Amos Tversky and Daniel Kahneman exposed almost 50 years ago:

1. We aren't as rational as we think we are: most of our daily behaviors are spontaneous, and even our deliberate decisions are biased.
2. One of the most powerful biases affecting our decision-making is our *loss aversion*.

What does that mean? We don't like the feeling of wasting our time and energy. So, when facing a choice between playing it safe (immediate, certain effort) or adopting an at-risk behavior (improbable and/or delayed pain) we might very well take our chances. Especially in situations and environments where such behaviors are more often than not *rewarded*: *I've done it before and didn't get hurt; everybody else does it; nobody seems to care that much…*

How are we to effectively influence a process that takes place in our minds in a fraction of a second, generally at a totally unconscious level? Well, it's precisely the leader's role – again, look first at yourself, then up, down and around – to try and tilt the balance by devising, deploying and living by the management practices that will support and demonstrate our total engagement towards our people's health and safety.

We can discuss leadership styles and the shape and form of those practices. But the bottom line is that you must constantly strive to enhance the values and beliefs related to health and safety, putting them at the very core of your organization's culture and ingraining them in everybody's hearts and minds. This is the only way to make everybody's behaviors – from engineers to managers, from supervisors to frontline workers – progressively safer. *One percent safer* with every action, with every conversation, every day.

Eduardo Blanco-Munoz is an EHS leader who has worked over 20 years in sectors such as chemicals, medical devices, power generation and aerospace. He holds a MEng in Health & Safety plus a MScEng and a MSc in Environment. He teaches BBS and Safety Culture at Université Sorbonne Paris Nord.

EVERY ACTION
EVERY CONVERSATION
EVERY DAY

TATTOO PIERCING OPEN

FROM THE OUTSIDE LOOKING IN

SIMON BLISS

Simon Bliss is Managing Director of Principal People and Founder of Safety4Good, which raises awareness of the positive contribution made by health & safety professionals by donating time to great causes pro-bono, offering a mentoring program for new and developing professionals, and fundraising for great charities.

I am not a safety professional. I sit outside as a supplier, supporting clients to find the best safety professionals for their organizations and supporting candidates to develop their careers.

As a supplier of the profession, I offer five observations from my experience that will help you achieve improved workplace safety performance:

1. In the past, people got hired on skills and experience and fired on behaviour. Make sure when hiring your lead Health and Safety professional the job AND person spec match your candidate. Make sure they have the right behaviours to fit your organization. Ideally hire a specialist safety recruiter to support your search (*"Of course I would say that" I hear you say*). However, the good recruiters will challenge you on the purpose of the role in order to ensure the role is clearly defined before the search begins.
2. Especially – as we learned from the pandemic in the time of COVID-19 – be prepared to consider and engage candidates from other sectors to your industry. Safety professionals have some exceptional transferrable skills and they will bring you fresh ideas.
3. Health and Safety professionals work in one of the few roles that touch all aspects of an organization. Other managers tend to work in silos wheras a good Health and Safety professional will get under the skin of the business and be an excellent temperature gauge on company culture, what's going on in the business, what's working and what's not.
4. Allow your safety leader and team to access the executive and board level: learning with be created in both directions.
5. Have a passionate board member take responsibility with the safety leader to partner their cause.

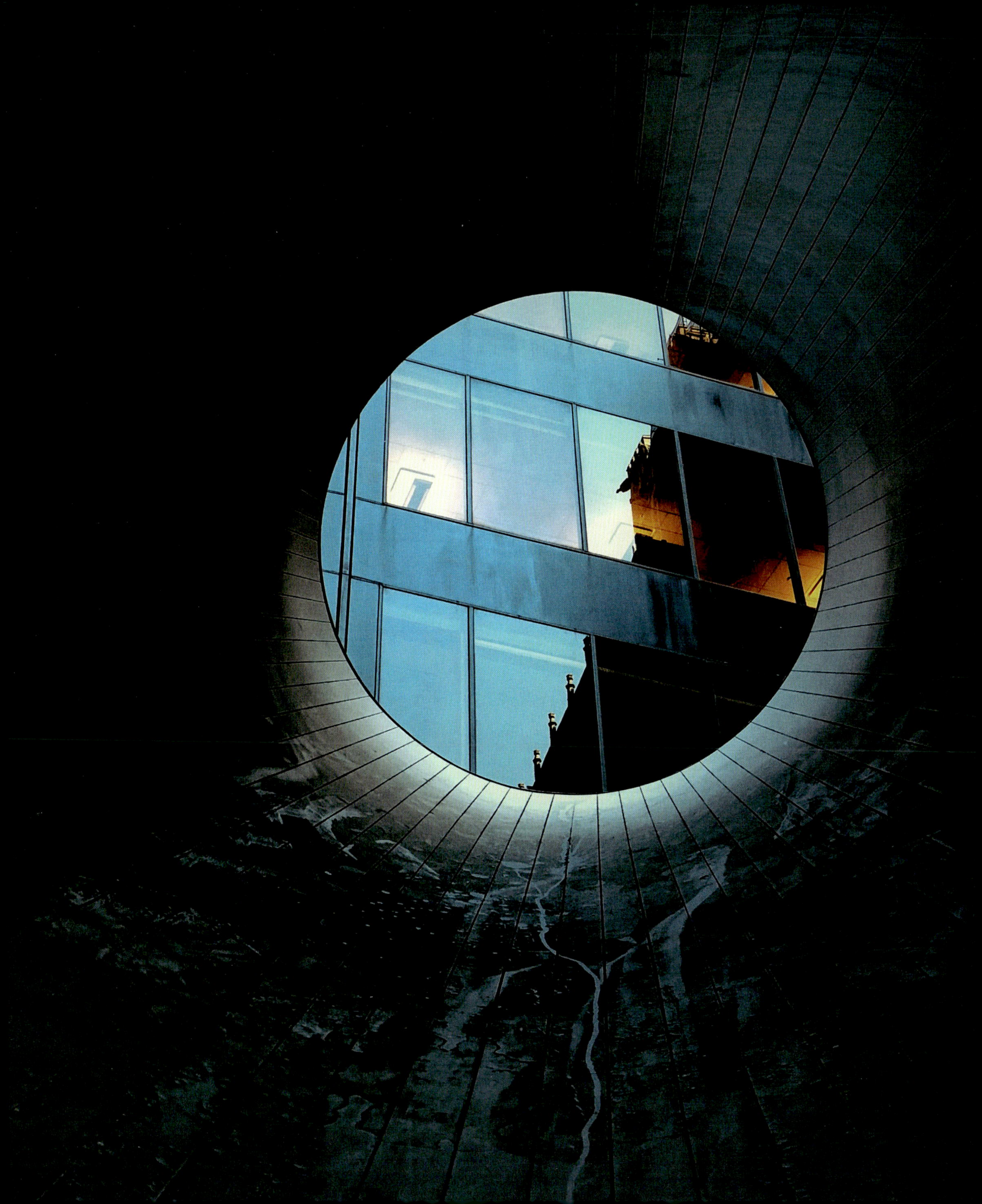

Image courtesy of NAPO Consortium

THE BATHTUB EFFECT

JOACHIM BREUER

Occupational safety and health institutions around the world are designing prevention campaigns with the aim of avoiding occupational accidents and diseases by drawing the attention to potentially hazardous situations at the workplace. When measuring the impact of such campaigns on the number of cases, in the first weeks and months, one will usually observe declining incident numbers. Once they have reached a low peak, there is the permanent risk, and high probability, of again increasing accidents numbers – the so-called *bathtub effect*. This is due to the fact that the attention raised by the prevention campaign starts to decrease steadily after its end. Studies on the effects of prevention campaigns clearly show that constant awareness raising is key for a sustainable reduction of cases.

This is where the European film project "Napo" comes in. The idea of the project is to promote safety and health at work through short videos. The videos highlight different risky situations during work, such as handling hazardous substances, the use of personal protective equipment or road safety. The main character, Napo, and his partners, convey in an entertaining and memorable way good practices for safe and healthy behaviour at work. It is part of the concept of the films to communicate occupational safety and health with a smile by also showing daily work situations during which not everyone acts flawless in every situation.

The described preventive measures are applicable to almost all work places in the economy, and as the characters express themselves in wordless language, everybody can understand the message regardless of any language skills.

By showing Napo films on suitable and regular occasions, such as during a Monday morning planning meeting with the team, the different safety topics can be brought to the attention of the staff. In combination with short discussion rounds, the clips will stay in mind and draw attention to the respective situation, which can appear sooner or later during the working day.

Napo films are a project of social security institutions from Austria, Italy, Germany, Switzerland, Netherlands, Poland, France and the European Agency for Safety and Health at Work. The initiative has already published over 30 Napo films and new videos are in production. In one of the latest films, Napo illustrates how to avoid the spreading of the virus in his company through suitable hygienic measures – a highly important topic in times of COVID-19.

As the comic style and the topics of the films are timeless and easy to understand, the videos can also be used as teaching material in schools and day care centres. Through periodically showing the clips in companies, schools, or even as pre-film advertisement in cinemas, we can make the world a little bit safer.

Napo films and learning material can be downloaded for free at www.napofilm.net

Dr Joachim Breuer is President of the International Social Security Association and holds a professorship for Insurance Medicine at the University of Lübeck. Breuer worked for over 30 years in various positions in the German Social Accident Insurance system, including as Director General, from which he retired in 2019.

'STOP THE JOB'... IS IT REALLY THAT SIMPLE?

TONY BROCK

Across the globe, many leaders exalt the mantra that if workers see something unsafe, they should raise the issue and "stop the job" to maintain a safe working environment.

But have we over-simplified this approach? Is it becoming a tick-the-box exercise?

Are we unintentionally abdicating responsibility to our work force and putting the onus on them to maintain a safe working environment, when it's really not as easy to 'Stop the job' as we imagine?

There is nothing wrong with the intent nor the commitment – most of us are authentic in giving our permission to stop the job to prevent an accident or injury. But the underlying premise is that it's within the power of the employee to stop the job, and there are many reasons why that may not be the case: physically or mentally, intentionally or unintentionally.

The workplace is complex: whether a research laboratory, factory floor, office complex or construction site. There are variables and daily pressures affecting staff. There are conflicting priorities and agendas, both personal and work-related, concerns of job retention and stress. These are further complicated by an increasing multinational and multicultural society. The willingness to stop the job depends on the confidence of the employee to speak up, and this is influenced by many factors: their level of experience, the relationship they have with their peers and supervisors, their own communication skills, and whether they fear retaliation for saying something unpopular.

It's nearly impossible to predict the variables that each worker will face in this situation. Safety management systems and tools do help greatly to provide consistent rules and procedures for the workforce to stay safe. But at the end of the day, it comes down to the individual themselves to follow the rules and assess the risks in their job. If they observe a risk, an unsafe condition, an unsafe act, we rely on them to speak up and stop the job. *And that's the key:* in order to stop the job, we need *them* to speak up. We need to create a safe environment for them to raise issues, to be heard, to be listened to and to be valued.

Although stop the job is a well-intended offer, it will only have a limited impact if we don't create an environment in which it is safe and comfortable to do so. When we create a safe environment for *everyone* to speak up, then softest spoken, the least experienced, the new person and the old hands will all be able to. That's a real game changer; it's a multiplier; it moves the dial on safety.

Have you ever stopped the job? Have you ever recognised a risk or an unsafe act and spoken up? Try to remember why you were compelled to do so.

How do you create a safe environment for individuals to speak up?

Do you have a culture that encourages speaking up?

How do you know?

Tony Brock is Head of Safety and Operational Risk at BP Exploration. Brock has over 30 years of global experience with BP in all over the world. His background is in Drilling, Production Operations, Upstream Engineering and HSSE.

WHAT'S RIGHT, NOT WHO'S RIGHT

MAKE IT EASY TO GET IT RIGHT

MARTIN BROMILEY

I've been privileged to strap myself to all sorts of airplanes and experience many things; to fly over 200 tons of machine close to the speed of sound, to make the horizon roll around at 400 degrees a second, to hang in my harness, deliberately allowing my trainee to fly temporarily out of control and recover. But my biggest privilege is to take responsibility for the safety of those who fly with me.

Through personal tragedy, I've learned even more about the science of safety, and especially that of human factors. In 2005, my wife died during an attempted hospital procedure. Issues of design, process and behaviours all came to the fore.

The science of human factors has probably been the biggest driver of safety in aviation in the last 70 years. It's also the only major science that's yet to have its day in healthcare. For me, it has a simple core message:

"Make it easy to get it right".

Design 'easy' into your systems.

Aviation has a proud tradition of designing everything based around the end user. Sadly, that's certainly not the same in healthcare and other industries. If you design systems, processes and procedures for others, understand from those that do the work how they *actually* do the work. Ask yourself whether you're making it easier for them or just harder to get it right. Understand "work as done" not "work as you imagine."

Behave in a way that makes it easier for those around you to be safe.

My leadership style is "open questions," followed by listening. I want to understand "what's right, not *who's* right." I thank those who speak up – they're trying to save my career, and possibly, my life. I never forget the accident reports I've read where someone knew what was going wrong but wasn't able to convey what they knew. Or wasn't listened to – just as happened to the nurses in my late wife's case. Those around you can also help identify threats, possible errors, and how to resolve them. Aviator's don't depend on "a wing and a prayer."

When things get tough, make it easy for YOU to do the right thing.

How much time do you spend mentally rehearsing and preparing yourself for high pressure moments? A colleague of mine talks about one traditional University in the UK that has the unofficial motto – "effortless superiority!" I can assure you, there's no such thing. The path to safety is laid on the foundations of effort and continual hard work.

Remember although you are a role model, in your own practices, you are not perfect. Do you role model humility and share your failings, do you demonstrate learning from those failings?

In my late wife's case, it was hard for the team to get it right, but it was remarkably easy for them to get it wrong. Elaine died. That day, the lives of seven clinicians were changed forever.

Martin Bromiley OBE FRCSEd. Bromiley is Founder of the Clinical Human Factors Group, a charity that has helped embed human factors in healthcare safety work. He's also a Training Captain for a major airline, and until recently, an aviation aerobatic instructor.

SAFETY WITH THE 'AND' IN MIND

URBAIN BRUYERE

How leaders react to bad news is the simplest way to assess organizational maturity:

Is the focus on *solving the problem and learning,* or on *finding faults and blaming*?

Does the term 'accountability' actually mean blame? In the real world, the answers to these difficult questions are nuanced, but more simply: do the behaviours of leaders encourage people to learn or to suppress bad news? This is a simple and effective way to assess organizational maturity and culture, and this approach works well beyond the safety arena.

Safety is a journey. There are no shortcuts, but there are boosters. Safety programs should be pitched at the maturity level of the organization to be effective.

Safety-I is about compliance and learning from incidents. Safety-II should be built on sound Safety-I foundations. Safety-II is about continuously learning and adapting.

Trying to bypass Safety-I and jumping straight to Safety-II does not work. Safety-II and the developments in human performance have made a tremendous difference in safety and productivity. Leadership and culture shape employee behaviours. In high-hazard industries, focusing only on leadership and culture is not enough. The critical barriers and controls are engineering-related and require sound process safety capability to be appropriately designed, maintained, inspected and continuously improved.

Deep learning from actual and potential high-severity incidents demands engineering and organizational psychology expertise. For these complex incidents, the initial phase of the investigation should focus on understanding what happened. What were the risk management failures that led to this incident? What barriers failed or were missing? The Bowtie diagram can be an effective tool for visualising these failures and understanding the strengths and weaknesses of the barriers that were supposed to protect individuals from hazards. The second phase of the investigation should focus on understanding why people did what they did. Why did it make sense to them at the time? What were the human, job and organizational factors that contributed to this incident? Disciplining the individuals involved without addressing the contextual factors leads to repeats. Issues will not be reported, rather, they will fester and cause even more serious incidents in the future. Deep learning only comes from understanding and addressing both the *what* and the *why* of incidents.

Safe production in high-hazard environments requires the integration of safety and engineering. Truly successful companies integrate not only safety and engineering but also quality, operations and other support functions. They demand safe quality production or no production. If we want our employees and organizations to thrive, we should simultaneously deliver safety with quality and productivity. Our challenge is to approach safety with the *'and'* in mind.

The response of frontline employees to COVID-19 has demonstrated that they can be incredibly adaptable and resourceful. Safety may involve additional complexity and cost. However, let us think beyond our safety silo and engage our frontline employees before bringing in additional bureaucracy and expenditure. We may be pleasantly surprised by the result.

Dr Urbain Bruyere is VP EHS at GSK. He earned a BP-sponsored doctorate at University of Pennsylvania, researching how teams improve safety and productivity by adapting to real-world complexity. Bruyere is a Fellow of the Institute of Chemical Engineers.

WHEN SAFETY-II MEETS BARRIER STRENGTH

CONTINUOUSLY LEARN AND IMPROVE

CONTINUOUSLY STRENGTHEN OUR BARRIERS

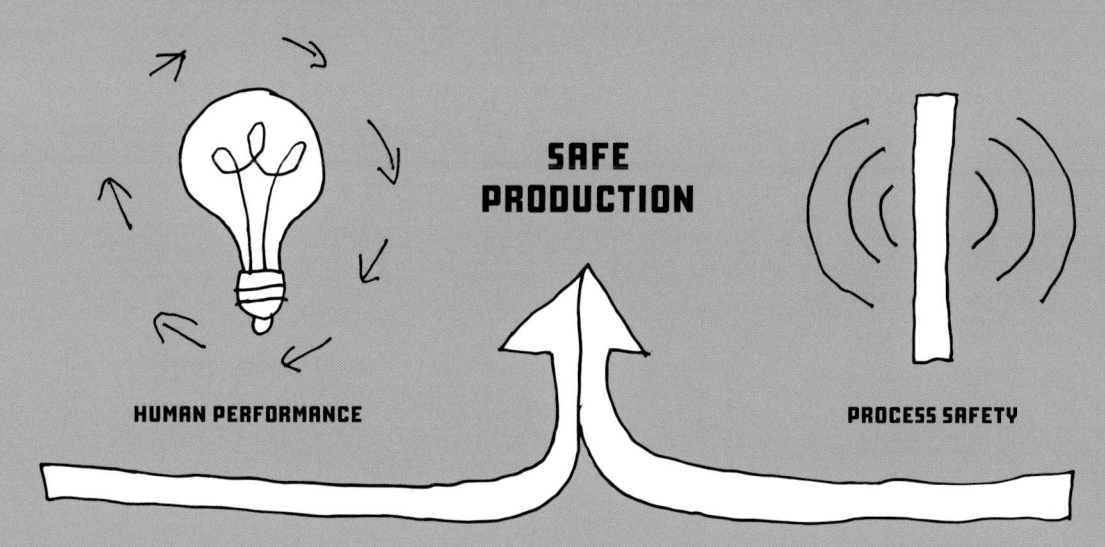

SAFE PRODUCTION

HUMAN PERFORMANCE

PROCESS SAFETY

ON BEING BOLD
RENÉ CARAYOL

"Everybody wishes they had hindsight, but it's much better to be practical, to be bold and a team player to solve the really big problems"

René Carayol is an internationally renowned keynote speaker and best-selling author in leadership and culture. He has been Chairman, CEO and MD of blue-chip businesses and worked with Fortune 500 CEOs and their executive teams. Carayol is a Visiting Professor at the Cass Business School in London.

EMBRACE THE PARADOX

TRISTAN CASEY

Uncertainty flows like water. Attempts to drain the water can be futile, as it simply channels its way to other entities. If the uncertainty becomes too deep, we drown in it, unable to function. But if we create a drought, we become bored and disengaged. Some uncertainty is helpful; it brings energy and engagement. Uncertainty can be clear, like a transparent and serene lake, or murky, like a rapid full of sediment and debris. Clear uncertainty is functional – we know the problem; we are just unsure about which path to take. Ambiguous uncertainty is dysfunctional – the problem is ill-defined, and the blur masks insidious hazards beneath its surface.

If there's one thing that the COVID pandemic has imbued us with, it's uncertainty.

Every way we turn, we're frozen with indecision and ambiguity. And it's more likely than not to be a defining feature of our future. Competitors are disrupting; climate change is looming; technologies are (re)invented faster than technical writers can keep up.

A lot of safety management is about reducing uncertainty, but do our current techniques simply mask or push the uncertainty to someone else? For instance, a principal contractor creates certainty by implementing blanket COVID rules: "No contractors from outside our state;" "All returning contractors must self-isolate for 14 days;" "No air travel, all contractors must drive to the site" are just some of the draconian rules I've heard.

The problem is that these 'certainty-creating' rules paradoxically create huge *uncertainty* for suppliers and contractors. They get 'stuck' with the uncertainty problem rather than it being jointly managed. Reducing risk in one area—public health—increases risk in other areas – like personal safety. Workers now drive 10 hours before getting to the site, rather than taking a 1.5 hour flight. Workers cannot see their families for months, as they can't leave site for quarantine restrictions.

Other certainty-creating safety activities lead us into a false sense of security. Creating prescriptive rules provides certainty in some situations, but the illusion of certainty in others—procedure isn't the best way to address the situation. The uncertainty is hidden by the dominance of certainty-creating processes.

Instead, the trick is to build capacities to cope with uncertainty. Explore new capabilities (embracing uncertainty) and exploit existing capabilities (to create certainty). Embrace uncertainty by being flexible (reconfigurable structures and systems) and reduce uncertainty by being stable (drawing on existing standards and processes). Go beyond the serendipitous: move away from chance combinations of the right people and the right time and towards proactive uncertainty detection and management.

Instead of trying to create certainty and focusing on risk, try focusing on uncertainty instead: What are you uncertain about? Why? Does the solution lie in existing capabilities and the creation of certainty, or should that uncertainty be embraced to stimulate innovation and new capabilities? The more we focus on uncertainty over risk, the more we are likely to improve safety and overall success.

Dr Tristan Casey is co-founder of research consultancy The Culture Effect and Lecturer at the Safety Science Innovation Lab at Griffith University. Casey's expertise includes leadership and culture, teamwork, training transfer and technology adoption. He is an endorsed organizational psychologist.

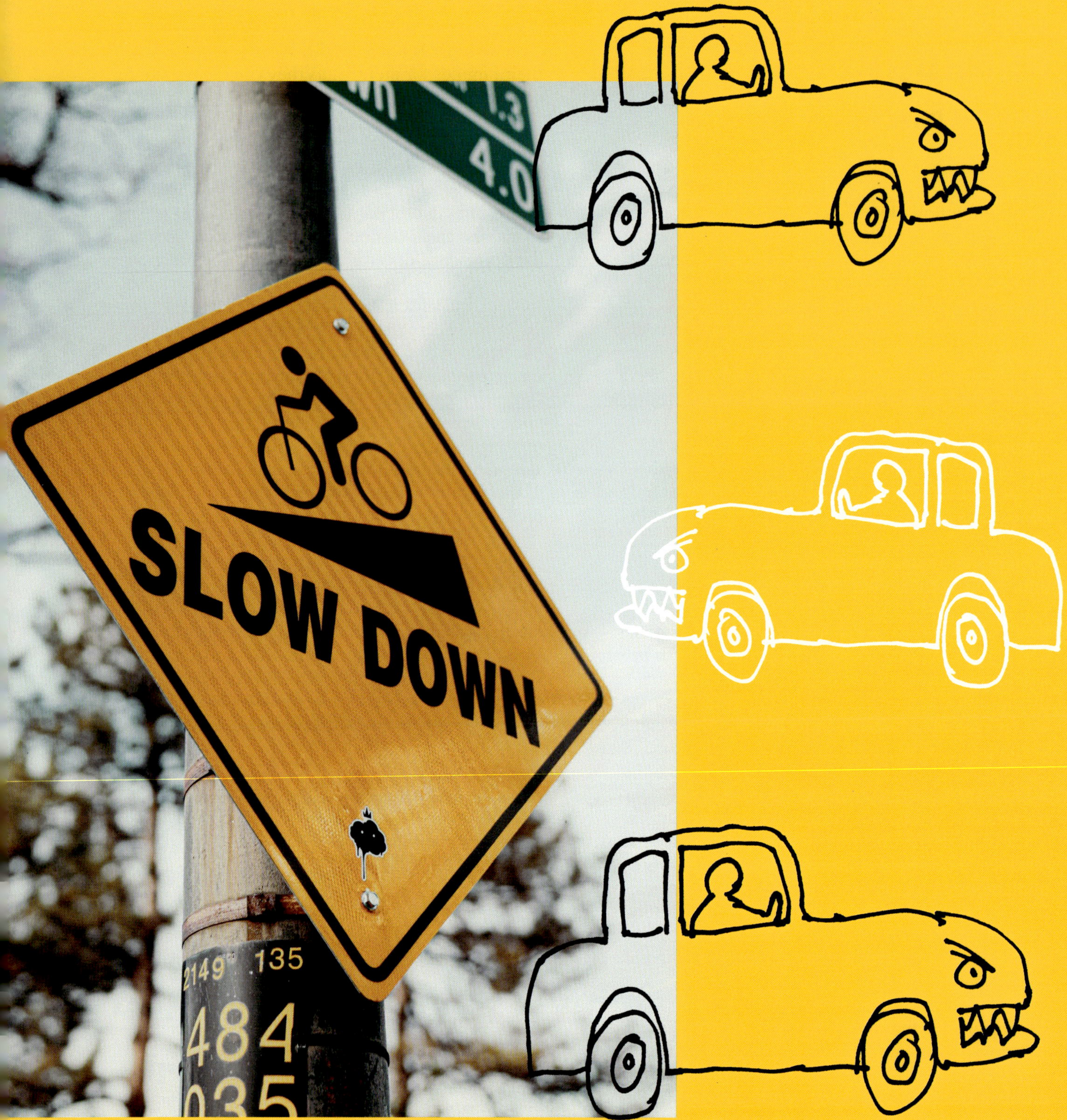
SLOW DOWN
484

PLEASE SLOW DOWN (IF YOU WISH TO SEE YOUR FAMILY AND FRIENDS TODAY)

EDMUND CHEONG PECK HUANG

Edmund Cheong Peck Huang is Chief Strategy and Transformation Officer at the Social Security Organization (SOCSO) of Malaysia. He's responsible for the organization's overall enterprise strategy and transformation initiatives. SOCSO is a statutory body governing the country's employment injury insurance, invalidity and unemployment and covering over 8 million workers.

Navigation applications – like *Waze, Google Maps, Garmin* and *TomTom* – as well as the use of virtual assistants like *Alexa* and *Siri*, are becoming increasingly popular with smartphone users worldwide. We could have these navigation apps or virtual assistants say to us: "Please slow down if you wish to see your family and friends today. You are arriving to a high fatality area in 500 meters."

Such messages can be changed, personalized and modified by sharing accident-prone areas, including statistics, and we can design these apps to say in many different languages and ways: "33 people said their last goodbyes this month in this area. Please stay safe," or even, "If you do not slow down, it's likely you will not arrive to your destination."

With information being at the core of our decision-making, maximizing the use of this information could be life-saving. Organizations collecting such data could share this in a centralized global or international database, and this information could be shared with the likes of *Google, Apple, Waze* or any new application developers who want to take part in a global initiative to make the world a safer place. Incorporating this data into applications we already use and making it available in new applications could make us more aware, and help us make better decisions, via safety 'nudges.'

If we can enable smartphones to give out these nudges, I believe we can achieve our target of making the world one percent safer.

MAINSTREAMING OSH INTO EDUCATION

LESTER CLARAVALL

Whether a worldwide pandemic, mass shooting, natural disaster, or serious workplace injury – one death is one too many when we are talking about precious lives. Often times, the result can be very devastating to communities, and the impact can be life-changing for families and friends suffering from the consequences.

Organizational and political leaders are tasked with ensuring the safety, health, and welfare of the people they serve, which should always be a top priority. But instead of solely relying on such people of power to keep us safe, the topic of safety and health should be everyone's responsibility. The world of work, the world of education, and the world of community should be collaborating and working together when it comes to safety and health for all.

Interestingly, many people go about their daily lives not really thinking about safety and health or just taking it for granted, especially in their working environments. This is especially true, given kids in the educational system don't really learn about the topic of safety and health in the classroom until they are out in the real world. Therefore, why not strive for efforts at an early age, so that as kids grow up, they are trained to have "safety eyes" that lead to "safety minds" when they are exposed to hazards?

In collaborating with the European Network Education and Training in Occupational Safety and Health (ENETOSH), the best way to reduce injuries is to provide lifelong training into safety and health from kindergarten to grade 12 (around age 18), known as the 'Whole School Approach'. Mainstreaming occupational safety and health (OSH) into education via curriculum policy will change the youth culture to one of injury prevention.

In the United States, Oklahoma became the first state in the nation to pass a "mainstreaming OSH into education" law (Senate Bill 262), which now requires the education department and the labor department to collaborate and provide workplace safety training to all schools in efforts to teach every student in grades 7 through 12 about safety and health. Senate Bill 262 uses the National Institute for Occupational Safety and Health (NIOSH) Talking Safety curriculum to educate students on the core competencies of workplace safety. The United States Department of Labor has since added safety and health as a recommended soft skill to prepare students before they enter into the workforce.

It's important to learn from the past in order to make improvements in the present. However, it's equally important to look into the future on what can be done today to make the world a safer place for tomorrow. Preparing now will help our children and grandchildren grow up prepared when they are faced with potential incidents that could hurt them or cause harm.

Lester Claravall is a public servant for the State of Oklahoma; he served two terms as national president of the Interstate Labor Standards Association. Claravall created the national award-winning youth employment game called Paying Attention Pays and the nationally-recognized video contest for teens called Speak Out for Workplace Safety.

LOOK INTO THE FUTURE ON WHAT CAN BE DONE TODAY TO MAKE THE WORLD A SAFER PLACE FOR TOMORROW

DEATH BY DISTRACTION

THEO COMPERNOLLE

'52 babies died in 2018, left in hot cars in parking lots throughout the US.'

Reading this headline, you certainly think: "How is that possible? They must be awful parents! That would never happen to me!" But you're wrong. It happens to the best of parents.

A young mother leaves home with her baby in her arms. She securely belts the baby in its car seat, and gets in the car to drop the baby at the nursery and drive to work. This is a routine which she executes with her *reflex-brain*, without really thinking. Her *thinking-brain* is still in the home zone, feeling grateful her husband is taking care of their sick toddler.

As soon as she starts driving, her phone rings. Her boss is angry: "The meeting with the board is at 10 a.m., and I am waiting for those figures from you!" She thinks "Gosh, I forgot," and says "I didn't realize you needed them before the meeting, I'll take care of it as soon as I arrive."

Her body goes into stress mode and her *reflex-brain* into survival mode – she can't afford to lose this job – while her *think-brain* switches 100% to her work context. She frantically tries to get in touch with the people she urgently needs this information from … and forgets to take the turn to the child center. She parks her car, gets out while still on the phone, rushes to her office and forgets a precious part that belongs to the home context: her baby.

Whether this will be just another anecdote or a tragedy only depends on the weather – or a watchful passerby. The sudden stress and the lack of sleep of a mother of a small baby increases the risk of making mistakes. Yet the major cause of these dangerous mistakes is: always being connected to our mobile, which compels us to multitask continuously.

The lethal combination of factors that lead to these tragic accidents also occurs in the workplace, if we don't pay attention to how human brains function.

Always being connected forces us to multitask with a *think-brain* that can only focus on one thing at time. Constantly switching between tasks causes continuous reorientations of our attention at the expense of our focus, thinking, understanding, remembering, creativity, calm and safety.

By using hands-free phones or touch screens while driving, hundreds of thousands of professionals, even safety managers, prove every single day that they do not have a clue about how their brain functions and to what extent their behavior puts themselves and others in danger. Considerable danger, as the risk of causing an accident is *8 times higher* when using a hands-free phone or voice command, and *23 times higher* when using touch screens.

By knowing how the human brain functions, and adapting our behavior to this knowledge, many accidents can be avoided.

Dr Theo Compernolle is an international consultant, trainer and keynote speaker. He has written several books including *Stress: Friend and Foe* and *Brainchains*. He has served as professor and lecturer at The Free University of Amsterdam and the business schools Solvay, Vlerick, TIAS, INSEAD, CEDEP and IMD.

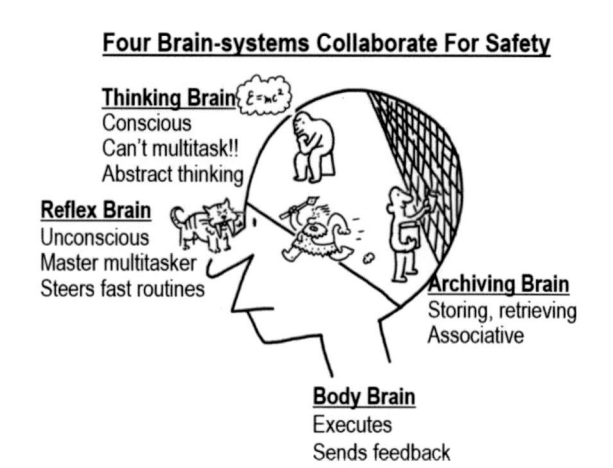

Four Brain-systems Collaborate For Safety

Thinking Brain — Conscious, Can't multitask!!, Abstract thinking

Reflex Brain — Unconscious, Master multitasker, Steers fast routines

Archiving Brain — Storing, retrieving, Associative

Body Brain — Executes, Sends feedback

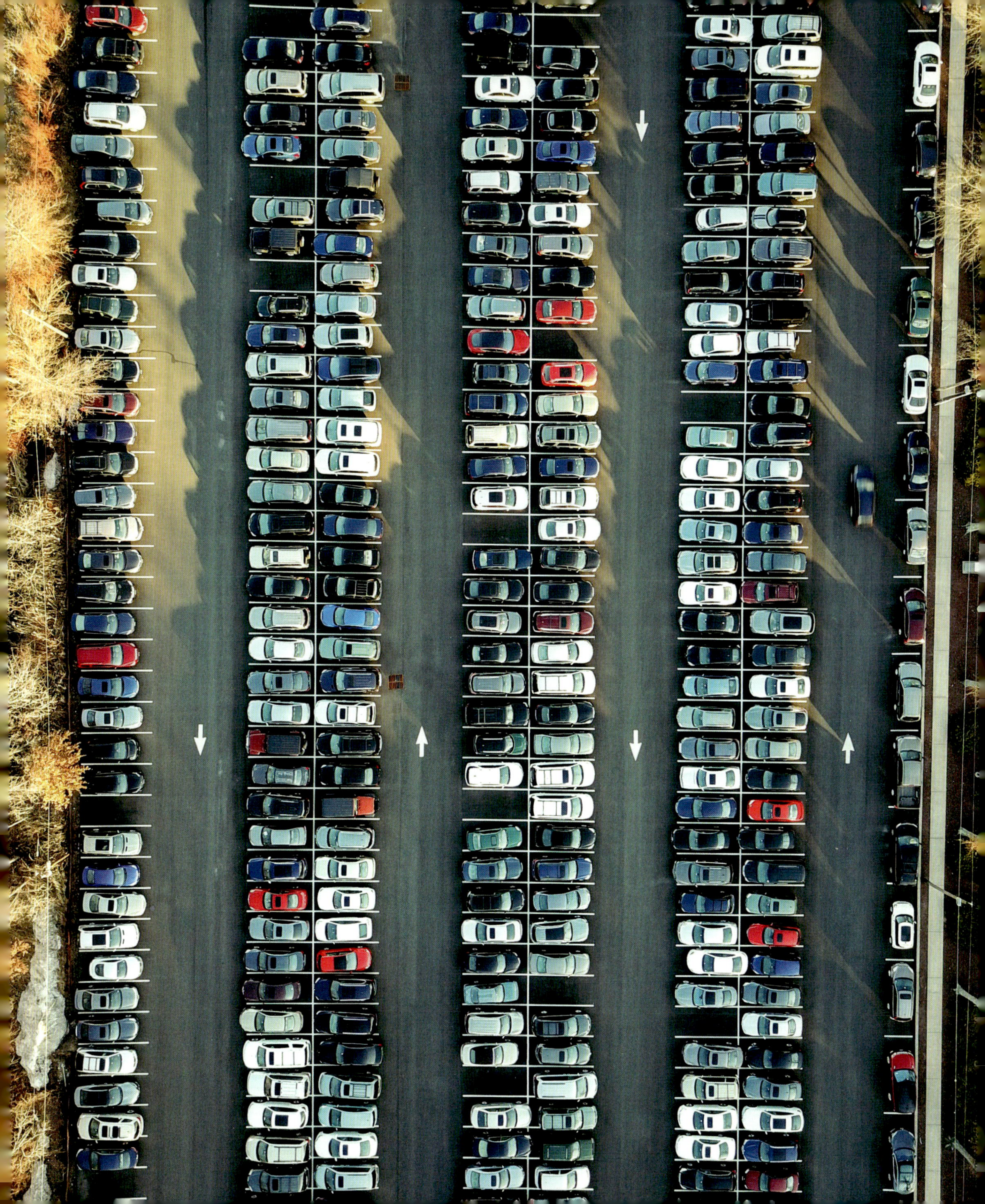

WORKERS ARE THE NOT THE PROBLEM TO BE FIXED, THEY ARE THE SOLUTION TO BE UNDERSTOOD

SAFETY REDEFINED

TODD CONKLIN

If you asked me what the biggest change in safety has been in the last 10 years, my answer is simple: *We have stopped seeing the workers as a problem to be fixed and started seeing our workers as problem solvers* – a source of important information from deep within our organization.

Our workers are not, and have never been, the safety problem. In reality, our workers come to work to do good work in a healthy, efficient and safe way. Our workforce is a group of the most knowledgeable experts on Earth in understanding how to snatch production and operational success from the jaws of insufficient and poorly-designed work management programs. Our workforce does this successfully the *vast majority* of the time.

The shift in the way we view our workers changes everything about the way we think about safety, and seeing workers as problem-solvers has forced our profession to change the way we define workplace safety. The old definition no longer is very informative or functional.

Safety is not the absence of accidents.
Safety is the presence of safeguards, controls, and the ability to recover.
In short, *safety is the presence of capacity*.

The long journey of workplace safety is not without some terrible moments of harm. Isn't it ironic that a profession based on keeping workers from getting injured has caused so much damage to the people safety intended to protect? For too long we have directed our safety efforts at the workers: we train workers; we observe workers; we track and trend workers. We have elaborate programs to manage and change worker behavior – all the while leaving the messy and complicated work management systems completely intact. We fixed the people and left the system alone. Then, we were surprised our problems did not go away.

What was wrong with us?

We held these beliefs, and we did it with the best intentions. Bad tactics with good intentions did not make for the improvement we desperately needed. The belief that bad safety was a result of a bad worker is a strong and comforting idea. It is much easier to believe the worker is bad and the system is good. Let's do a small logic exercise: If safety is the absence of an accident and bad workers cause accidents, therefore bad safety must be a result of bad workers.

And then there was some good news, shifting philosophies and some startling revelations. We learned there were more levers we could pull to create change than just worker behavior. Slowly, and with much resistance, a new set of levers appeared. This new way of viewing our workers began to become more acceptable. Now, organizations all over the globe have revitalized and improved safety and reliability not by changing their safety programs, but by changing the way the organization's leadership thinks about the workforce. Workers are the not the problem to be fixed, they are the solution to be understood.

Dr Todd Conklin spent 25 years as Senior Advisor for Organizational and Safety Culture at Los Alamos National Laboratory, where he has served in the Human Resource and Reliability Management and Training arena for the last 15 years.

THE GREATEST THREAT OF ALL

SAM CONNIFF

It could be said that there's one defining and overarching health and safety brief to humanity, which is to leave things in better shape for the next generation. For the first time in several lifetimes, we're the leaders who are about to fail that test.

It has been my lifelong experience that too many of the rules, norms and behaviours we find ourselves sticking to, day in day out, are simply bad habits and hangovers that have become entrenched over time. Eventually, they become the rules that represent the risk, and breaking them (or -remaking them) is the only responsible act. They may have worked well in a past context, but in consistently failing to overhaul them, we risk failing the next generation.

The vast improvements in health and safety practices have, over the last fifty years, undoubtedly saved millions of lives and businesses by protecting welfare and livelihoods from bad practice and mismanagement. Our workplace environments are safer than ever before, but as a species, we are not.

Governments, scientists, business leaders, economists, and the UN, are all in unison that the greatest threat to global stability, progress, and our very existence is the climate and ecological crisis. Yet somehow, the relationship between businesses cutting carbon emissions and governments' ability to deliver policy that will reverse our current trajectory, falters again and again.

It strikes me that here there is an opportunity for the H&S industry and wider community to step up and step into the vacuum of leadership on climate change. To write some new rules and set new standards, which will force business and government to U-turn at the speed required. The invitation is clear: to reconsider writing rules that uphold human health, safety and wellbeing, by looking not at what is, but *what is to come*. If the traditional power in policy work isn't working, what could a rigorous new set of ISO frameworks do to protect the lives and livelihoods at near-term risk from the ecological emergency?

In the short term, we will face more weather that will make it impossible to work outside. In the longer term, an ice-free arctic summer will seriously increase the speed of global warming and endanger the lives of millions. The usual levers of control and protection are failing our future; we need new health and safety regulations that have the power to inflict serious penalties on businesses or states that continue to carry out life-threatening practices in relation to the climate crisis.

The brief is bold and urgent, but the H&S movement is uniquely positioned to respond. It is time *now* to repeat the successes of the last half century in protecting lives and livelihoods, but with an honest and ambitious eye on the next 50 years. It's time to broaden the scope, widen the aperture and redefine what health and safety means in its fullest context against a real risk assessment of the greatest threat we all face.

Sam Conniff is an entrepreneur and author, co-founder of LivityUK, DigifyAfrica and Don't Panic. He is also the author of *Be More Pirate,* called "a magnificent book" by *Forbes* and named Business Book of the Month by *The Financial Times*.

HAPPY (AND HEALTHY) AT WORK
CARY COOPER

Sir Cary Cooper, CBE, is the 50th Anniversary Professor of Organizational Psychology & Health at Alliance Manchester Business School, University of Manchester. Cooper is President of the Chartered Institute of Personnel and Development (CIPD), President of the Institute of Welfare and Immediate Past President of British Academy of Management.

Even prior to the COVID-19 health crisis, many organizations globally were looking at creating a healthier workplace by making organizational wellbeing a strategic imperative. I chair the National Forum for Health and Wellbeing at Work, comprised of 40 C-suite leaders of global organizations such as Rolls Royce, BP, Shell, GSK, Microsoft, UK Government, NHS Employers, BBC and John Lewis Partnership, so I hear from a variety of large employers on these initiatives. One of the key issues in creating health and wellbeing strategically in the workplace is ensuring that we have more line managers with *emotional intelligence*, that is, social and interpersonal skill of empathy and compassion.

We need parity between people with technical skills and 'people' people. Too often people are promoted into managerial roles based almost solely on their technical skills, and if we want to improve the health and wellbeing of the workforce, we need workers to be managed by people who care about them and support them in their jobs. Managers with emotional intelligence support workers' autonomy, fashion and encourage manageable workloads and realistic deadlines. These managers enable workers to have good work-life balance. If we promote this type of line manager from shop floor to top floor, rather than 1% safer, we would make the workplace 10% safer and healthier.

As John Ruskin, the great social reformer, wrote in 1851: "In order that people may be happy in their work, these three things are needed: they must be fit for it, they must not do too much of it, and they must have a sense of success in it."

THE NUMBER ONE PRIORITY

DOMINIC COOPER

"Safe production is the number one priority" is a philosophy that places safety and production on an equal footing, while removing ambiguity and providing guidance to all.

In other words: "Hey guys, we have to produce, but we have to do it safely."

The underlying rationale is that the greater the compatibility between safety and production demands, the safer people will work. It recognizes that safety is an operations responsibility with the leadership of each business unit acting to help employees realize an injury-free environment.

"Safe production is the number one priority" is also much more believable to employees than vision statements that promote "safety is the number one priority" or "safety is a value," when people's everyday experience actually shows that production is.

A vision statement that encapsulates this and senior management's intentions in relation to developing a safety culture should be constructed in such a way that it actually guides the in-depth formulation and implementation of their strategic safety culture plan, e.g., what the strategic direction is and how the organization is going to achieve it.

An example might be:

Safe production will be understood to be, and accepted as, our number one priority by everyone in our organization. In our continued efforts to achieve a sustainable level of zero incidents and be recognized as achieving world-class status, we shall use safety culture principles to identify and demonstrate implementable, measurable and sustainable solutions, with measurements showing continuous improvement.

The 7 Safety Culture Principles that will guide our actions going forward, are:

1. As a company, we will put people's safety before profits.
2. All of our employees are encouraged to speak up about any safety issue, at any time, so actions can be taken to fix them.
3. We shall adopt servant leadership as our preferred style throughout the organization and lead by asking: "What can we as leaders do for you, to help you do your job safely?"
4. We will construct all of our risk controls as a team: employees, leadership and OSH professionals together.
5. We shall endeavour to ensure all of our safety communications are two-way dialogues between people.
6. We shall try to ensure that we train people in safety until they cannot get things wrong, not just until they get it right.
7. We shall encourage everyone to report every near-miss and incident and endeavour to learn as much as possible from each one to try to prevent recurrence.

Using these seven safety culture principles will help companies create a safety partnership between management and the workforce, wherein they jointly work toward the achievement of common and understood safety goals, with clear and consistent communication, efficient monitoring, reporting and feedback, resulting in decisive action to investigate blockages and taking the appropriate corrective action as needed.

Dr Dominic Cooper, CFIOSH CPsychol, has served as a Professor of Safety and Industrial & Organizational Psychology. Cooper is an authority in behavioral science and safety culture. An award-winning author, he's written books on behavioral safety, safety culture change, safety leadership, and motivation, and over 200 professional and scientific articles.

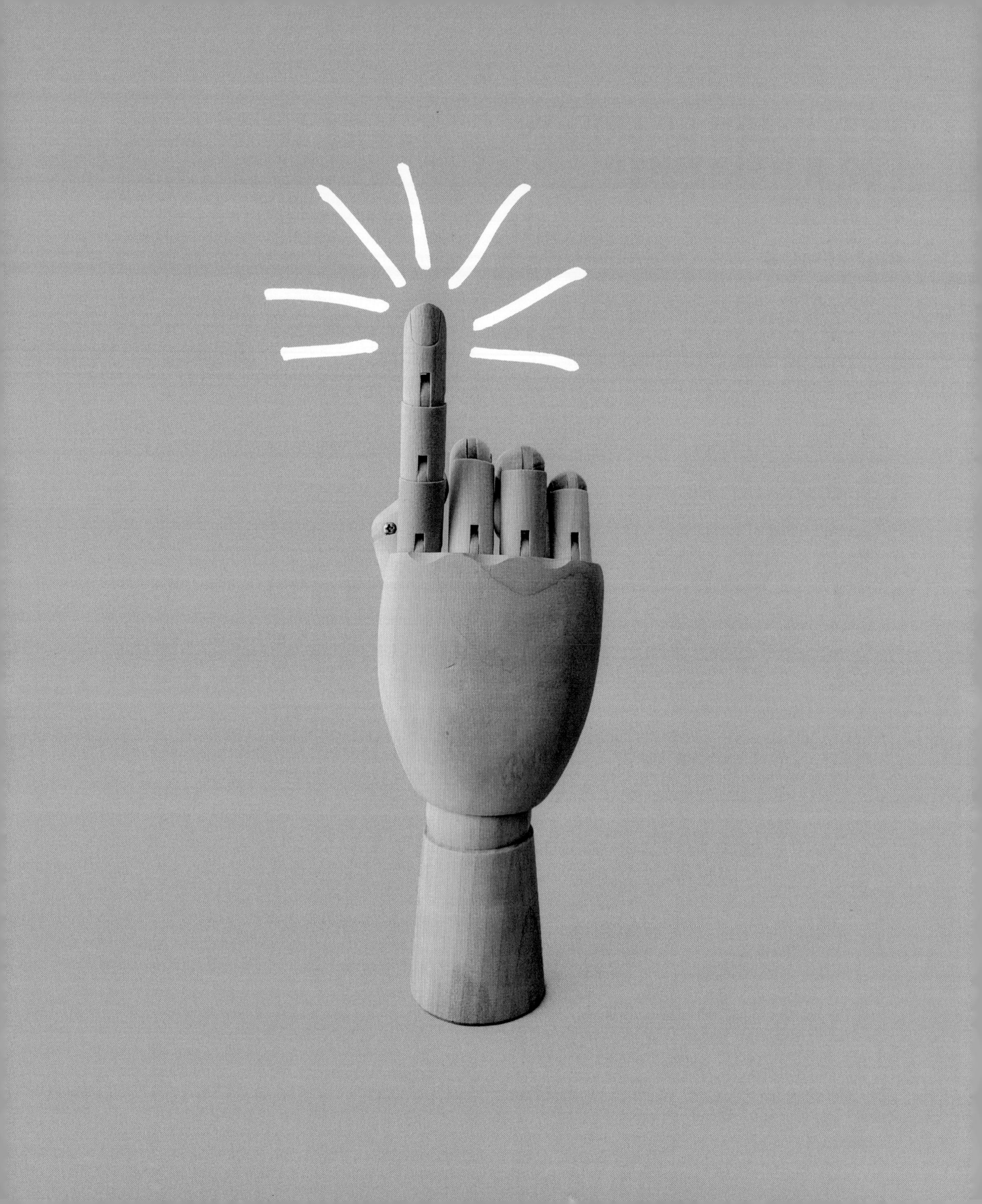

EVOLUTION OF SAFETY FOR A HEALTHIER PLANET

MIKE COSMAN

As my career in health and safety evolved over 40+ years, so has the time horizon over which I view change. As a frontline HSE inspector in the 1980s, I looked to identify the quick wins that might make a difference immediately, often focusing on acute risks in construction, machinery or transport. In the 1990s as a sector leader in Health Services and Construction, I was mainly concerned with the planning cycle of 1-3 years. It was time to introduce some supply chain initiatives and look at some longer-term health risks.

In recent years in New Zealand, I've been fortunate to have the opportunity to think about more fundamental changes at an industry or national level in forestry. In 2014, we embarked on the reform of the whole health and safety framework of legislation and regulations.

Now, my target is governance and organizational strategy. This is where the really big decisions are made and where the real power and influence comes from. We all know the saying 'what interests my boss fascinates me.' Well, take that up a level and think about investor and share/stakeholder interests and how these impact the boards of our public and private sector organizations.

The Environmental, Social and Governance (ESG) criteria increasingly being used by investors to differentiate between companies are incredibly powerful, because they align purpose, people and performance. Work done by Dr Kirstin Ferguson and others has demonstrated a statistically significant link between ESG reporting and health and safety performance.[1]

ESG reporting now increasingly includes health and safety at various levels including supply chain standards, modern slavery, decent work, etc. Alignment with environmental and climate change goals increases the impact. So, if I want to think about reducing exposure to carcinogens, my focus will be much less on the individual workplace and more on creating a narrative that links together elements of the ESG agenda. What is the environmental footprint of these chemicals? What are the conditions for those who produce them? What are the downstream implications for my customers of using and disposing of this product? What are the implications for my brand of being associated with a harmful manufacturing process?

In New Zealand, after the Pike River Mine disaster that highlighted fundamental governance failures, we introduced a proactive due diligence duty on 'officers' to demonstrate strategic leadership, oversight and risk management. This duty extends to the risk of harm to anyone who may be impacted by the conduct of the business or undertaking – much wider than employees and contractors or even up and downstream duties. In a landmark case, a horticultural marketing company that set growing standards for its fruit suppliers was held liable following a fatal accident.

Now I spend most of my time influencing directors, challenging their norms and perceptions – and hopefully shifting the dial 1% or more in pursuit of the cleaner, greener, safer, more sustainable future we aspire to in New Zealand.

1. https://medium.com/carbonclick/what-is-esg-and-why-is-it-important-f9036bb96d66

Mike Cosman is a partner in CosmanParkes Ltd., a strategic health and safety consultancy based in Wellington, NZ. He is Chair of the New Zealand Institute of Safety Management, a Chartered Member of IOSH and NZISM and a Member of the Institute of Directors (NZ).

TO SYSTEMIZE OR TO HUMANIZE?

ANDI CSONTOS

As leaders in health and safety, our impact hinges on our willingness and ability to challenge the status quo, to be compassionate, curious and courageous, to ask better questions and seek smarter answers:

> *How do we build a better, healthier and safer working world?*
>
> *How do we move beyond the ideology of 'Zero Harm,' towards an ethos of 'Plus One' and positive contribution?*
>
> *What is our core purpose as health and safety professionals?*

The world has become (and will continue to become) more volatile, uncertain and ambiguous. The context within which we are searching for answers and trying to effect change is increasingly complex. In attempting to disrupt, reinvent and refresh, we have lost direction and become disconnected. We have built empires around safety, defined safety as a process and not as an outcome of work, establishing cumbersome safety management systems that self-perpetuate our own busy-ness. We declare a new language for Safety II and demand new safety systems.

However, our purpose was never to do 'safety' or to do it differently.

Our challenge was, and remains, to work differently.

To focus our efforts on designing meaningful, efficient, safe and productive ways of working. To strive for integration and simplification of systems, removing layers of complexity through worker-led problem solving and innovation. To improve job roles and task design and directly impact human performance; to reduce fatigue, stress and cognitive load. Doing these things will lead to healthier and safer outcomes.

Our challenge is to humanize.

To acknowledge diversity, personality and creativity – the inherent nature and individual differences that make us human. To recognise that cognitive performance and our ability to judge risk, make effective decisions and undertake work tasks safely, requires us to maintain high levels of physical and mental health. To know, that without a healthy workforce, we cannot achieve desired business outcomes, be they competitive advantage, innovation, productivity… or safety. Putting health, resilience and capability at the very centre of business strategy will lead to positive contribution, for our people, our organization, our society; and a better working world.

Andi Csontos is a Partner with Ernst & Young Australia with over 20 years of health and safety experience and a background in psychology. Csontos is regularly sought after as a strategic thinker and change agent; delivering business and safety outcomes through engaging people in problem solving and work design.

Ask better questions

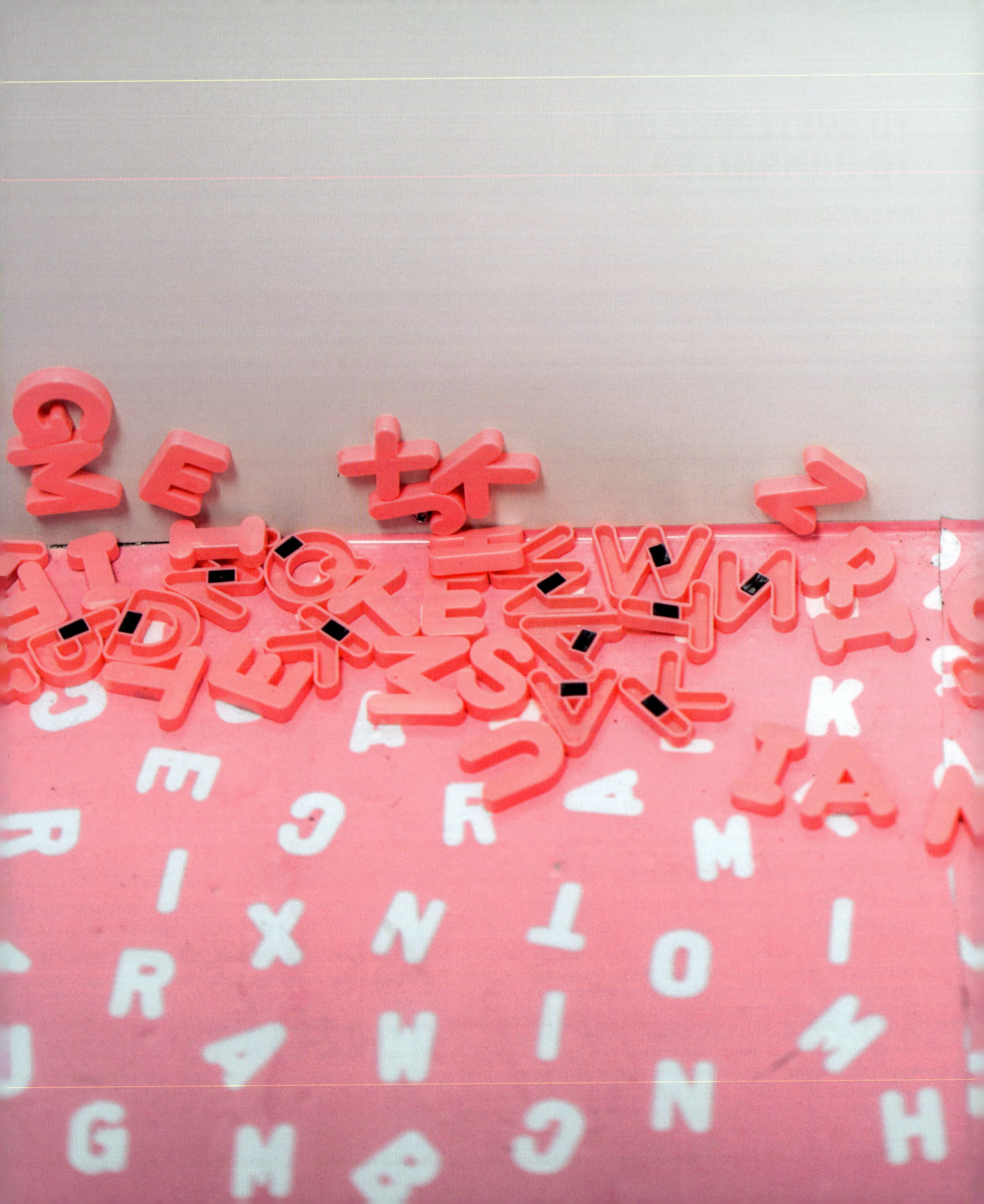

BILINGUAL SAFETY

ANNE DAVIES

The language of safety is more than just words. It is about context and understanding.

I grew up in a bilingual household. I was educated through the mediums of Welsh and English. I spent my school years translating sections of textbooks from English into Welsh. But it was not a literal translation; it was an interpretation. I had to understand the true meaning of the words and their context.

I encourage leaders and practitioners to behave 'bilingually'.

Translate the business language to the safety team. Translate the safety language to the business team.

Be confident in your pursuit, to truly understand each other's words, their context and their meaning.

In so doing, create the "roots and wings" to develop and grow healthily and safely, making your place of work "just a little bit safer every day."

Anne Davies is a business crime specialist and partner in GunnerCooke LLP. She has handled regulatory investigations for 30 years, assisting companies and individuals to minimize risks through criminal and civil liabilities arising from internal or external investigations. Davies contributed to *Corporate Liability: Work Related Deaths & Criminal Prosecutions.*

MENTAL HEALTH IN TRAUMATIC TIMES

SHAUN DAVIS

"Trauma" is defined by the Oxford English Dictionary as "a powerful shock that may have long lasting effects," and it can relate to a wide range of situations and circumstances that present a real or perceived threat or danger to us as individuals. In the aftermath of our safety being compromised, we experience trauma.

Many of us can relate to the feeling in connection with the COVID-19 pandemic.

Between 7% and 14% of people will experience a traumatic event at some point in their lives.[1] Furthermore, about 10% of these people will develop trauma-related illnesses and are likely to have longer-term symptoms that would benefit from professional help and intervention. The response to trauma depends on a number of factors, including previous exposure to trauma, how it was managed and the support networks available. Our personality, emotional resilience and life experiences, as well as our level of self-awareness and ability to express feelings, will dictate how we react and recover.

Fortunately, there are positive actions that support mental health in the aftermath of a traumatic event:

- Accept that things take time: It is important to acknowledge that it will take time to recover from the experience, but if you're still dealing with the effects of the incident after one month, seek medical help.
- Keep people close to you: In the wake of a traumatic incident, it's tempting to keep to yourself and withdraw from social situations, but it is important to stay connected and keep people, especially those that you can open up to, close to you, whether physically or virtually, as is possible.
- Maintain your routines: Keeping to a regular routine will help to get you back on track. This includes sleep times, meal times and trying to get back to work.
- Stay on the straight and narrow: Although it can be tempting to drink excessive alcohol or turn to recreational drugs to help numb the memories in the aftermath of a traumatic incident, these can actually intensify your symptoms and "self-medicating" with them should be avoided.
- Recognize that what you are feeling is perfectly normal: Regardless of how you are feeling in the aftermath of a trauma – whether your feelings and emotions are more intense and unpredictable than usual, or if you don't feel different at all – this is perfectly normal. We all react differently and need different types of support, information and reassurance to help overcome trauma.
- Take time to relax and refresh: Keeping up with hobbies and pastimes will assist recovery, as will finding the time to relax and unwind. This will allow your mind and body to recover from the experience they've been through.

Ultimately, it is important to remember that trauma can be a life-changing experience, and whilst it can be meaningful and significant at the time, most people do recover. It is important to be patient with yourself.

1. Rick J, O'Regan S, Kinder A (November 2006) *Early Intervention following trauma – a controlled longitudinal study at Royal Mail Group*, Institute of Employment Studies, Report 435.

Dr Shaun Davis, MA, MBA, MA, MSc, MA, LLM, CDir, FIoD, CFIOSH, FCIPD, FIIRSM, FICA. As a Chartered Director, Davis has a wide portfolio of responsibilities across the Royal Mail Group, including safety, health and wellbeing, alongside regulatory compliance and security. A published author, Davis prides himself on his "pracademic" approach of combining practice and academia.

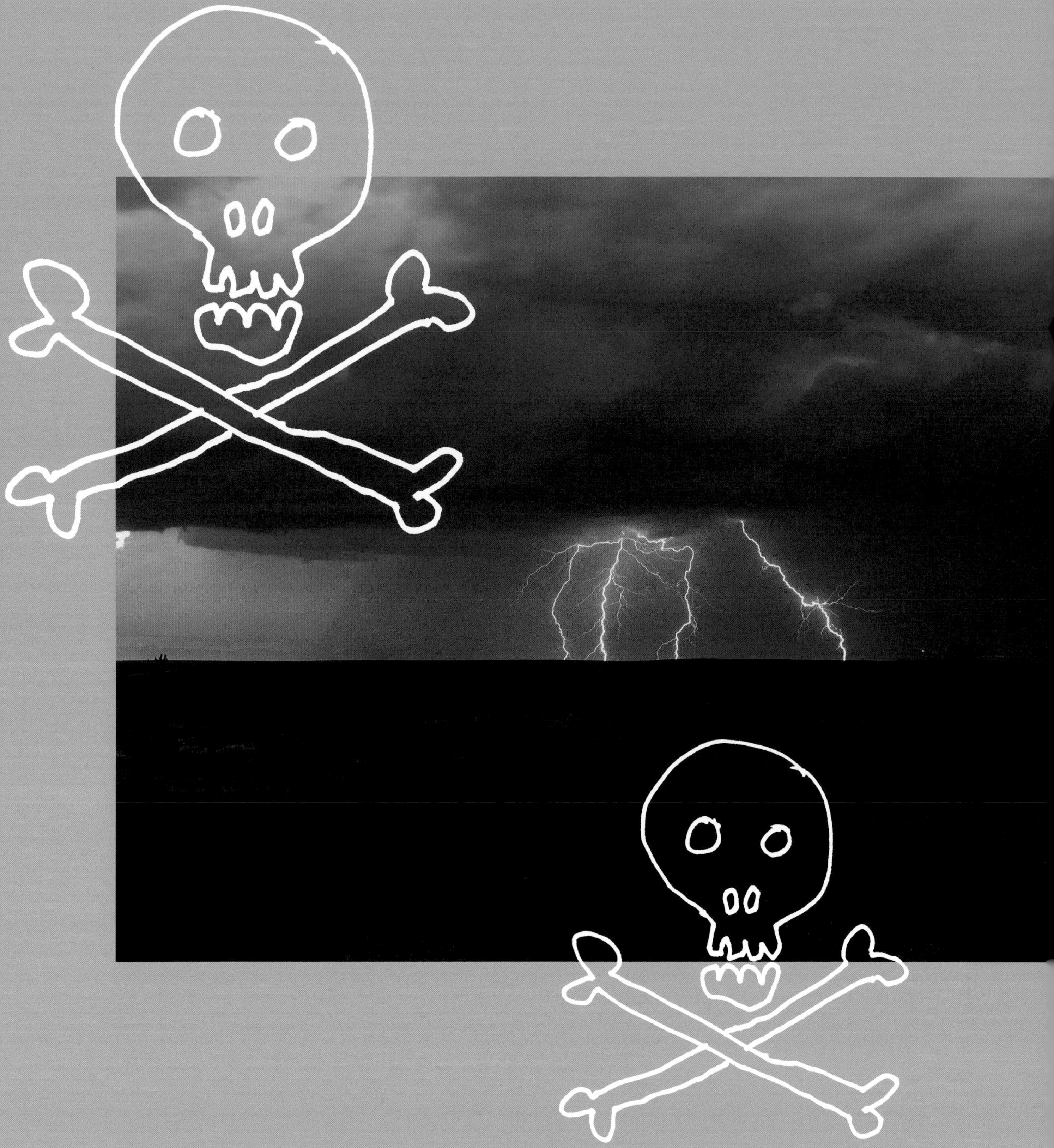

SUNSHINE AND RAINBOWS

KIMBERLEY DE SELINCOURT

Perhaps I've seen one too many NHS rainbows while out on my daily lockdown exercise, but I can't help thinking that if we adopted a more 'sunshine and rainbows' mentality, safety would improve. Love and kindness will save the day.

Love might seem tangential from occupational safety, but the two are intertwined. After all, if you have love, you have something to lose. When in love, suddenly getting home safely from work is more compelling: you have someone relying on and waiting for you.

Equally, to employ the old 'carrot and the stick' approach, if love will not drive people, then perhaps fear will. Growing up in a rural area, 'the dangers of the farm' were drilled into us from a young age. In one harrowing educational video in particular, children played in a grain silo and drowned after losing consciousness from the toxic gases contained within. Twenty-five years later, this type of tragedy remains prevalent in confined spaces the world over. Much like COVID-19 was underestimated, if we're unable to perceive the level of risk in a situation – as is the case with the 'invisible threats' of hazardous atmospheres and diseases alike – then we give little gravitas to its danger.

'Seeing is believing' they say, but toxic environments aren't a supernatural phenomenon up for debate: they are real and will kill you.

These are sides of the same coin. Be motivated to come home safely for loved ones. Conversely, do not run into confined spaces unprotected to save friends.

What if we could harness the energies of love and fear into safety training events?

In training, harnessing powerful emotions – whether positive or negative – is key to instilling safe behaviours and reducing accidents. With compelling, emotive training sessions, workers consider the real-life ramifications of their actions and feel more drawn to look after each other as though they were family.

Since learning is both mood-state and context dependent, rather than conducting classroom training, run sessions in environments where workers are likely to experience hazards. If possible, recreate elements of the physiological stress that would be experienced, such as through the use of a simulator or VR training. Help them imagine the risk and the personal ramifications that would be faced.

Strong emotions forge lasting memories. Carry out an exercise whereby workers list off their favourite pastimes – listening to music, enjoying the outdoors – and then go through how they would feel if they could not enjoy these activities should they lose their hearing, their eyesight, or a limb. Ask what factors stop them from being safe, like uncomfortable PPE, and ask them to consider the impact on their families should the worst happen. Then have them to weigh the two options.

Engaging workers in emotive training sessions, in which they gain an understanding of the real dangers and ramifications of their actions is paramount to reducing workplace injuries and deaths. Harnessing emotions can foster a safer culture and more days ending with 'sunshine and rainbows.'

Kimberley de Selincourt is the Editor of *Health & Safety International* magazine, *Health & Safety Middle East* magazine and *Air Water Environment International* magazine.

FOSTER A SAFER CULTURE

WE DON'T NEED MORE RULES

SIDNEY DEKKER

(cartoon: "Zero Harm" badge next to a poster titled "OUR 100 101 LIFESAVING RULES". Two figures; one says, "Well – how are we going to get 1% safer otherwise?" Signed "By Sidney Dekker")

Dr Sidney Dekker is Professor at Griffith University in Australia, where he founded the Safety Science Innovation Lab, and Honorary Professor at the School of Psychology at The University of Queensland. He's the author of 11 books on human error and safety culture.

COMFORTABLY NUMB
PHILIPPE DELQUIÉ

You are cruising down the wide-open highway, smoothly, confidently: the driving conditions are perfect – dry road, great visibility, light traffic – and by now, you are comfortably over the speed limit, with an unquestioned feeling of safety. Now visualize this: your right front tire suddenly blows out.

Picture the scenario in vivid details: the load popping sound, shreds of rubber flying in a flash and hitting all over, the car beginning to swerve, while the steering wheel spins madly between your hands; all this in the same fraction of a second!

Are you ready for this? Would the mere thought of this scenario ease your foot off the accelerator a bit?

It is well-established in psychometrics that we tend to get habituated to the current level of a stimulus (ambient sound level, temperature, your pace on the highway, etc.). The same goes for risk level: we can become comfortably numb to it. If you have slightly escalated the risk in your position and everything continues to be good, then there will be no incentive to roll it back. What-if thought experiments can be a powerful, inexpensive strategy to break out of mental numbness, and regain situational awareness that may have slipped away.

Thus, from time to time, engage in the mental effort of bringing up to mind possible hazards and threats, small and large, which may have become conveniently shrouded. This would be like an occasional recalibration of your internal, virtual risk-o-meter.

This gives a whole new meaning to the famous claim that "only the paranoid survive," doesn't it?

Dr Philippe Delquié is a professor of Decision Sciences at the George Washington University. His expertise and scholarly studies are in Human & Machine Decision Making and Risk, spanning their statistical and psychological foundations.

SHARK
SIGHTED
TODAY

ENTER WATER
AT OWN
RISK

IS THE RISK WORTH IT?

RUTH DENYER

This, for me, has always been the question that speaks to the essence of what we do; it's the reason that the role of the OSH professional is so fascinating and exciting.

It's a deceptively simple question, but one that is incredibly hard to answer. Perception of risk, the attitude of the players, broader social norms, and the purpose of the activity, will all color the decision. This five-word question hides a nuanced and complex decision-making challenge, and when I reflect on it there are a few key principles that have governed my response:

- This question isn't for me to answer. My value is in providing a reasoned framework to support the decision making
- The people who own the activity need to own and understand the risks (upside and downside)
- My role is building the full picture, in a way that people understand and on which they can base an answer.

What's the risk worth?

Activities are undertaken because people are striving for the best outcomes. This will mean different things in different organizations and to different individuals. As a rule, people set out to achieve success so, any risks they are willing to take will be framed within this. To really get people to think about what risks they are willing to take in order to achieve their version of success in a given situation, you have to get under the skin of their challenges. 'Why' is as important as 'how.' I love to spend time really understanding *what* this success is, because then I know why people are taking these risks to achieve it. I also learn so much more about what motivates people to make the decisions they do, which then leads to, simply, better conversations.

Try asking – 'What does success look like?' and listening to the answers.

How do we understand and quantify risk?

We have often framed the downside of health and safety risks too strongly in terms of legal compliance. In my experience, people don't want to harm or hurt others emotionally or physically, and perhaps a focus on legal requirements and compliance removes the human compassion from the evaluation and real consideration of the risk level. The human story is so strong, and it should lead any consideration of the impact of a risk.

Risk in its entirety can never be eliminated. Our role as leaders is to effectively support those making these hard calls, providing them the fullest possible picture to base these decisions on. Ignorance is our greatest enemy: a wilful or inadvertent overlooking of the factors involved prevents a balanced assessment of whether, given the context, the outcome justifies the risk.

Answering the question requires the courage to face all factors honestly.

So: *Is the risk worth it?*

Ruth Denyer, BSc (Hons), CMIOSH, MIIRSM is a risk professional with health and safety expertise and is the Group Operational Risk Director at British TV channel ITV. Denyer promotes key principles of the 'new view' of safety, driving a culture of risk ownership within a holistic risk framework.

Encompass everything

SEE THE BIGGER PICTURE

BERNIE DOYLE

I learnt early on in my managerial career that trying to be "everything to everybody all of the time" was a recipe for failure. Therefore, whenever possible, I have aimed for a high level of engagement with all staff members of every organization I have been involved with. This applies to every aspect of the business – which of course includes safety.

Sharing more with the staff always resulted in higher levels of engagement, which in turn, led to increased results in every section of the business.

In every business I had the fortune to be involved in, I would attempt to remove the "invisible" barriers that existed between staff and non-staff members. One business had staff members not required to account for their time, while the majority of the other were required to "clock in and out." Within a very short time, we moved away from a two-tier system to a single method: everybody had to account for their time. Beyond parity, the other upside was that now I had a system I could use as a head count in the event of an emergency.

In regard to the safety of the staff, I made the shop floor team totally responsible for this, as this is their work area and they, more than anybody, knew the risks.

However, I did not abandon them. Rather, I offered *more* support – this would allow them full responsibility to stop any machine that they thought was unsafe. At first, they were unsure about this new level of power, but as expected, they treated it with respect and used it wisely. Once I had convinced them that they owned the shop floor, the change was dramatic. Housekeeping reached new levels, and with support, they introduced new checklists for machines and surrounding areas.

When we hosted visits from our customers, I encouraged team members to lead the visitors around and explain the workings, in particular, how they dealt with changes in production, brought on by the customers changing deliveries at the last moment.

This occurred in an environment that initially had one of the worst records for injuries, according to our government agencies, and they were surprised at the level of improvement that occurred in a relatively short time, just six months.

It was clear to me that while the safety improvements were outstanding, we had to reach for even higher goals. This occurred when we introduced 'Lean Manufacturing' – this discipline reinforced the basis of teamwork is everything. Furthermore, it forces the shift from a top-down to a bottom-up approach, in which, if followed correctly, ensures everybody works for the team, and therefore adds value.

Within 12 months, we were acknowledged with an award for 'Outstanding Safety Improvement" and from our customers, additional business as we were able to become the best supplier, moving from the bottom of their rating list to the top.

It's not always just about safety, and you can't pick on one element of the business and ignore the others – you need an approach that encompasses everything.

Bernie Doyle is President and Chairman of the National Safety Council of Australia Foundation (NSCA) and Honourable Secretary General of Asian Pacific Occupational Safety & Health Organization (APOSHO). He served as Senior Executive in several organizations. For much of his career, he worked in the automotive industry, including over 20 years with General Motors.

PROMOTING COLLECTIVE ACTION

OCKERT DUPPER

About 60 percent of global trade today is organized within so-called global supply chains (GSCs). While the contribution of GSCs to global economic growth and job opportunities is evident, their impact on the living and working conditions and the safety and health of workers in developing countries raises important concerns. The International Labour Organization (ILO) has noted that failures at all levels within GSCs have contributed to decent work deficits in a variety of areas, including occupational safety and health.

Paradoxically, however, supply chain relations can also create opportunities to ameliorate these effects and contribute to supporting improvements in arrangements and outcomes for safety and health for workers. While current efforts to improve working conditions in GSCs (such as private compliance initiatives) have resulted in some improvements, these have been episodic, unstable and uneven because they often fail to address the root causes of non-compliances. It has become clear that in order to address the root causes of the most serious and entrenched OSH deficits in global supply chains, a new paradigm is required – one that involves the collective action, influence and resources of all major stakeholders in the supply chains.

In 2015, the G7 established the Vision Zero Fund (VZF). Administered by the ILO, the VZF aims to eliminate work-related deaths, injuries and diseases in global supply chains around the world. The main objective of the Fund is to improve OSH practices and conditions in sectors that link to GSCs, and to strengthen institutional frameworks, including labour inspectorates and employment injury insurance schemes in countries linked to such GSCs. Currently, VZF is operational in eight countries across three continents and in two supply chains, namely garment/textiles and agriculture. It works primarily in low-income countries, and a pre-condition for funding is the commitment of countries and stakeholders to prevention and the implementation of minimum labour, environmental and safety standards. To date, activities have directly benefitted almost 40,000 government officials, employers, workers and their organizations, ultimately improving the safety and health of an estimated 2.5 million workers.

To achieve its main objective, the VZF implements a strategy entitled "collective action for safe and healthy supply chains." This strategy is defined as:

> *"A multi-stakeholder approach that involves governments, workers and trade unions, employers (both national, transnational and global) and their organizations, multilateral organizations, civil society and development agencies, working together so that each meets its responsibilities consistent with organizational roles, to implement an agreed plan or set of actions to reduce severe or fatal work accidents, injuries or diseases in global supply chains."*

VZF operates under the principle that it is only when all relevant stakeholders assume responsibility that the root causes of OSH deficits in GSCs can be addressed in an effective and sustainable way. All VZF county project teams use ILO's convening power to bring these stakeholders together to design, develop and –most importantly – implement industry-wide strategies.

Please join the VZF to truly make global supply chains safer and healthier for everyone.

Professor Ockert Dupper is the Global Programme Manager of the Vision Zero Fund (VZF) for the International Labour Organization (ILO).

CRONOS
bulker
CRONOS
bulker
CRONOS
CRONOS
CRONOS
bulker
CRONOS

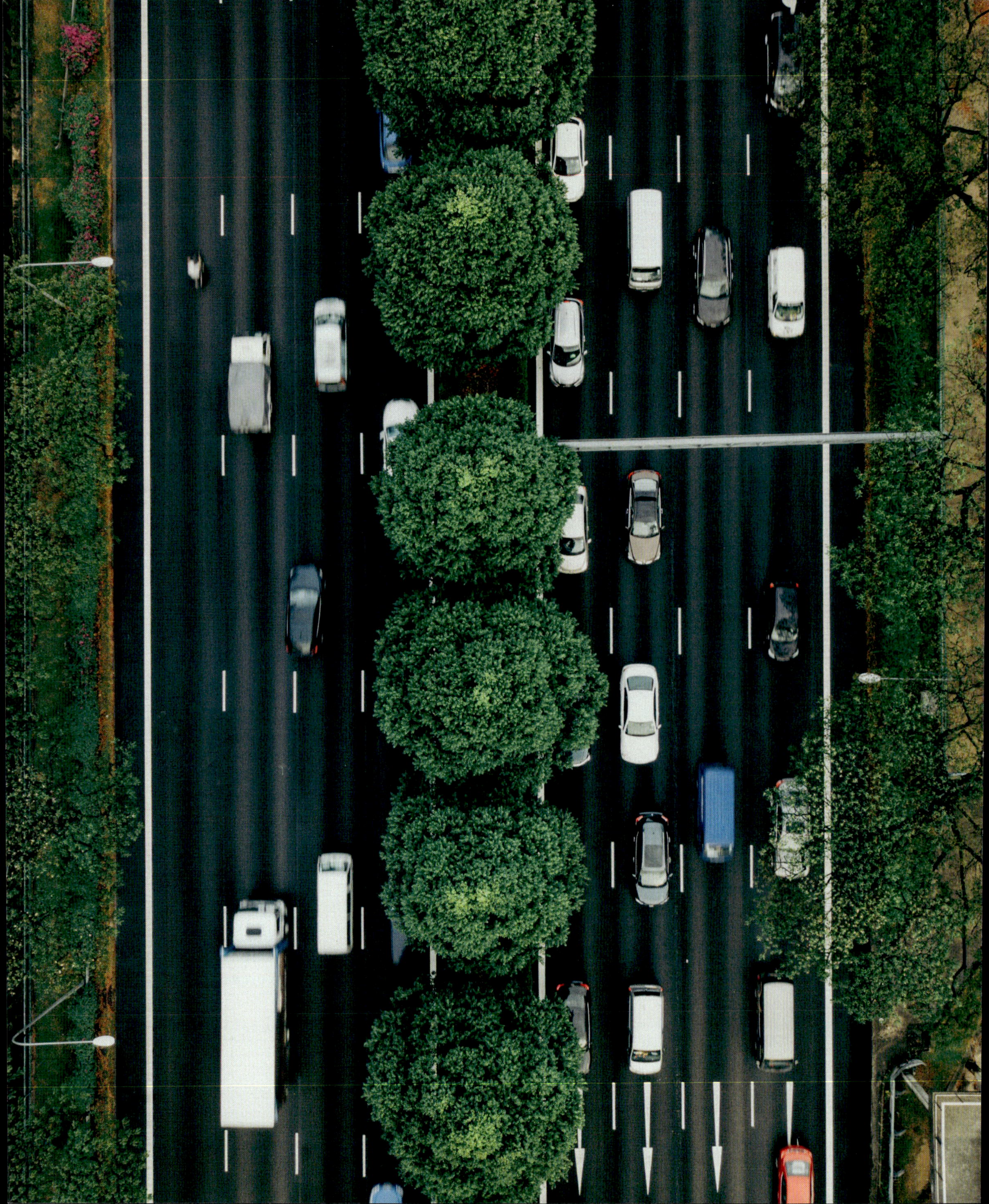

TRAVEL SAFELY

WALTER EICHENDORF

Risk assessment is, or course, obligatory within the EU and many other countries around the world. Looking into the risk assessment of hundreds of companies in many countries, I realized that two of the main sources of fatal and severe accidents are usually not covered, these are:

Commuting accidents

Commuting is a leading cause of work-related accidents for many companies around the world.

In other words: The journey to the workplace can be the most hazardous part of the workday, because in many professions the risk of commuting accidents is higher than that of workplace accidents.

Nevertheless, large parts of the safety and health organizations within and outside companies do not take care of them. Typically, they are not included in the risk assessment. However, resources to avoid commuting accidents are available, and OSH Wiki offers a good summary under "Commuting Accidents."

National organizations – like the German Road Safety Council (DVR), provide risk assessment tools that are easy to adapt to the needs of companies. Check the equivalent organizations in your country for local versions.

Duty travel

Whilst this situation is quite similar to commuting accidents, there remains a need to help organizations address their safety, health and security responsibilities towards workers (including employees, contractors and volunteers) when travelling for work or on international assignments – and these assessments may need to include the dependents of your workers, too.

As with commuting accidents, risk assessment tools that are easy to adapt to the needs of companies are available. The International SOS Foundation provides an excellent example, which is called "A global framework: Safety, health and security for work-related international travel and assignments."

Risk assessment is one of the most powerful tools to ensure workplace safety and health. Including commuting accidents and duty travel in the risk assessment will save hundreds of lives.

Professor Dr Walter Eichendorf is president of the German Road Safety Council (DVR) and professor at Bonn-Rhine-Sieg University of Applied Sciences. His focus is on a 'Culture of Prevention'. He serves on national and international advisory boards and as a member of the board of directors, European Transport Safety Council (ETSC).

MARGINAL GAINS ALL ADD UP

TIM ELDRIDGE

Ever since the concept of 'marginal gains' was introduced into our collective consciousness by Sir Dave Brailsford, explaining the approach to making GB Cycling so successful at the Beijing Olympics in 2008, I have been convinced it has a tangible beneficial application to the world of occupational safety and health.

The notion that you can break down everything you do into micro-processes and then improve each one by 1% can absolutely be applied to the world of work. If this, as a minimum, makes your workplace one percent safer then you have achieved something as significant as saving a son, daughter, mum, dad, husband or wife from losing the one thing that is most precious to them.

So, what does a marginal gain in the safety of a workplace look like? This is where 'Nudge Theory' comes into its own.

We know people at work generally do not deliberately do things unsafely, but rather, it is often a product of their subconscious. So how can we 'nudge' this subconscious behaviour into being marginally safer?

A really obvious application of nudge theory was during the coronavirus pandemic. Simply placing lines that are 2 metres apart on the floor outside a supermarket, for example, is 'nudging' us to behave safely.

My challenge to business leaders is simple – what nudges can you apply to your work environment that will deliver a marginal gain – that, added up, can make our world of work at least one percent safer and healthier?

A couple of simple examples of 'nudges' could be:

- Marking the floor around work equipment with footprints to show where the operator is safe to stand
- Moving office printers to an area away from desks so employees have to get up and move around to access them (and maybe also encouraging them to print less to support your environmental goals)

Small nudges – when added up – have the power to push us in a considerably safer direction.

Tim Eldridge has over 25 years' experience managing health and safety. He has led health and safety functions for large global organizations and OHS regulators in both the private and public sectors, in the UK, Middle East and Australia. Tim is a Vice-President and a Chartered Member of IOSH.

THE POWER TO PUSH US

THE HUMAN CONDITION OF THE ORGANIZATION

JAKE ESMAN

Equipment and safety procedures are important. But they have limits.

Life is simply more complex.

What happens when we acknowledge the boundaries of hand-holding protocols and protective measures?

Well, we are forced to look beyond the mechanics of safety and consider the human condition of the organization. And this is an opportunity to expand our approach to safety by fostering self-awareness, self-reliance and empowerment.

The best equipment and gold-standard protocols support safety. Leadership focused on the human condition of the organization deepens the culture of safety through connection and people care.

So, how do you, as a leader, impact the human condition of your organization to create a culture of safety beyond procedures and hand-holding?

Jake Esman works at *Bridge – the way over*, an international leadership advisory firm. He works with executives and leadership teams on organizational development questions with particular focus on spirituality in leadership. He lives on the Isle of Wight in the south of England and is a cold-water swimmer.

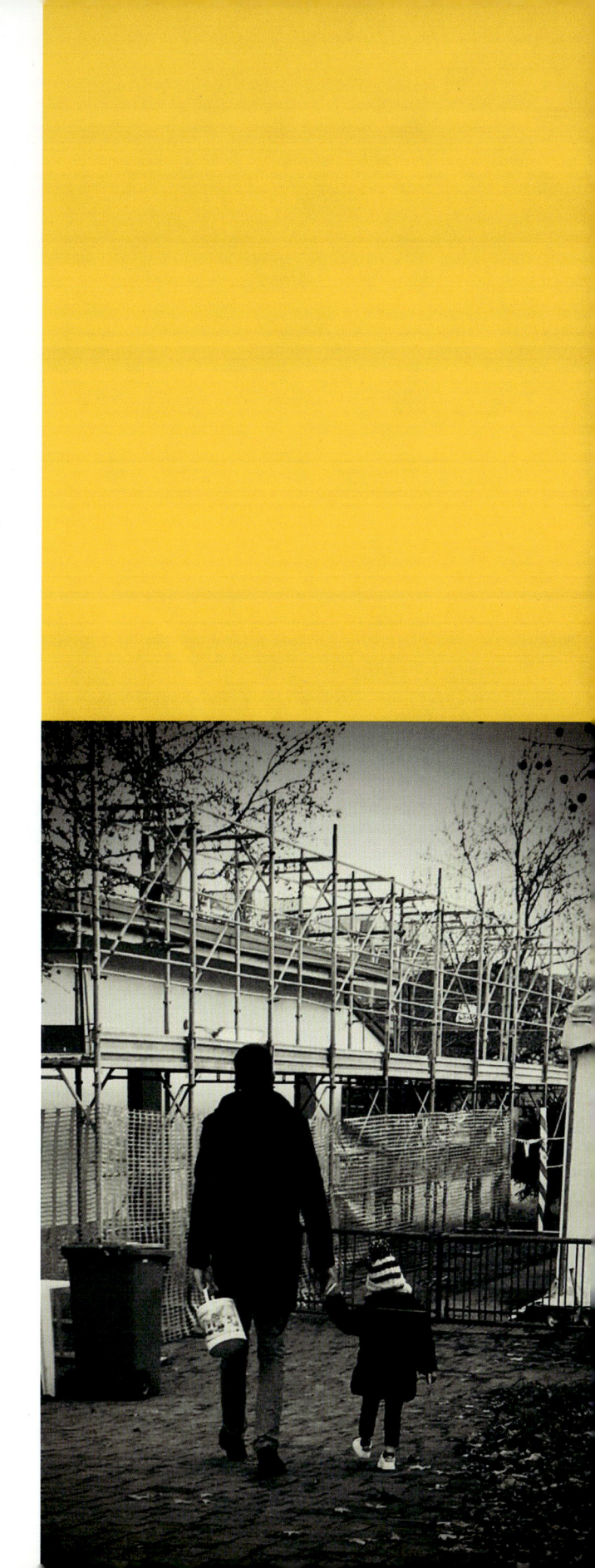

and breathe

BREATHING EASIER AND SAFER

GRIGORII FAINBURG

The 2020 coronavirus pandemic has led humanity to reflect on the safety of air quality and airflows – a consideration that hasn't normally been widespread. Any air – regulated and visible or unregulated and invisible – is a dynamically active mixture of chemical compounds, their clusters and aerosols, including electrically charged ones.

Some of these aerosols contain pathogenic bacteria, viruses and coronaviruses and can be called 'bioaerosols'. The size of bioaerosols is various – from invisible to dust – and they can *easily* penetrate human lungs. Protection against them and the effects of their contamination has now become a central problem for the security of humankind. Our global security requires bio-air safety.

Different workplaces and air environments require different technical means and methods to manage bio-air safety. Ensuring the safety of the air environment is technically very complex and systemic. It's a special culture and is affected by the deviation of human behavior. Bio-air safety measures are aimed not only at protecting against bioaerosols, but also at maintaining high indoor air quality in working rooms and transport cabins, which need adequate protection measures. All this will require the development of standards for the organization of bio-air safety culture. It's a global process. It means cardinal changing of air ventilation, new engineering solutions, time and money.

Unfortunately, this complex engineering challenge often cannot be fully met within companies. The other path to protection is to reduce the exposure effectiveness by activating the protective forces of the body itself, seeking to support homeostasis and immunity.

We have some existing methods of treatment and healing for chronic COPD, dust bronchitis, asthma, rhinitis and pollinosis and other respiratory diseases (approved by Russian health authorities and the accumulated experience of clinical observations and research). These methods, addressing health of the air and the body, can intersect. One such way is the use of 'air-potash salt rocks interaction', the use of potassium-rich rock salt to generate nanosized particles and small negative ions.

Speleotherapy (subterranean respiratory therapy inside caves and mines) has been used in potash mines of Russia (since 1975) and Belarus (since 1990). Since 1989, sylvinite speleoclimatic cameras, known as 'vital air rooms' or 'salty air rooms', have demonstrated good respiratory results for tens thousands of adults and children. All employers connected with workers' lung and allergy occupational diseases (pneumoconiosis, silicosis, asthma, dermatitis) might implement these methods in practice, perhaps in the form of post-shift rehabilitation.

These methods may also be useful for the control of COVID-19 or other coronaviruses at the stage of initial infection (suppression in the air, nose and throat) and for the rehabilitation of recovering patients who have damaged lungs. Of course, scientific research on this coronavirus is in its infancy. But our knowledge about aerosols and lung diseases like dust bronchitis, pollen allergies and influenza, shows that there is promise.

The application of these new drug-free rehabilitation and revitalization methods can help prevention of lung diseases and rehabilitation for patients in recovery. Leaders of organizations that encounter bio-aerosol risks should take note of the research and fashion a future where workers breathe easier and safer.

Professor Dr Grigorii Fainburg is an Honored Higher Education Worker of the Russian Federation, Full Cavalier of the "Miner's Glory," Director at The Institute for Safety and Health of the Perm National Research Polytechnic University. He is the author of legislative acts and interstate standards, textbooks, curricula, monographs and patents on safety and health.

ON BEING EMPATHETIC

KRISTIN FERGUSON

Imagine a world where everyone – whatever their formal role – chose to lead with empathy.

Empathy doesn't cost money, doesn't require a policy and doesn't need supervision. You can't see, touch or pop empathy into a box. But you can definitely feel empathy when you receive it, making empathy a powerful tool to earn trust and strengthen relationships in the workplace, both essential ingredients to building a safer environment for all.

In a nutshell, empathy is your ability to put yourself in someone else's shoes. Having empathy means you are self-aware enough to put ego aside and tune into other people and their feelings and really listen to what they have to say. Sometimes leading with empathy means being able to sense what someone may be feeling before they can even use words to describe whatever it might be.

I can hear people now reading this and thinking: "That sounds great but, in my workplace, we just tell it how it is and don't have time for all this airy-fairy bullshit. How can having empathy possibly keep people safe?"

Leading with empathy does not make you a weak leader or one that focuses only on emotions and not the operational needs of your business. Leading with empathy means the way you achieve operational excellence, safe production or whatever other goal you might have is achieved in a way that brings people along with you rather than distrusting whatever you might have to say. Leading with empathy makes you a leader people choose to follow because they know you have their best interests at heart.

Imagine if leaders were able to sense that their team members felt unable to speak up (about a safety issue or anything else)? That leader might then be able to ask constructive, open-ended questions to uncover the real issue lying beneath the surface. This, in turn, will build trust within the team and a culture much more likely to see people willing to raise concerns, stop work when a situation is unsafe or suggest process improvements that remove risks.

Leaders with empathy take responsibility to understand the individual situation for every person they lead. They know, for example, that one team member has just experienced the breakdown of their marriage or the loss of their partner's job. They understand that this particular team member may be feeling distracted or anxious and perhaps not as focused on the job. If you lead with empathy, you can help this individual and avoid what might have become a tragic (and avoidable) workplace incident.

Empathy is the cornerstone of being an emotionally intelligent leader. The more leaders we have in the world leading in this way, the safer we will all be.

Dr Kirstin Ferguson is a business leader, author and an award-winning expert in safety governance and leadership. You can connect with Dr. Ferguson at www.kirstinferguson.com.

> EMPATHY DOESN'T COST MONEY

HIT PAUSE

RHONA FLIN

It is easy to see in hindsight when a task could have been slowed down or paused in order to check the risks, and thus avoid an accident. While workers are usually encouraged to "stop the job" if they have safety concerns, in practice, this does not always occur.

A powerful illustration of the need to pause a job is shown in an animation[2] of an accident on an offshore drilling rig, the Maersk Interceptor, in 2017 (all credit to the company for sharing this valuable lesson). A heavy seawater pump was being lifted on a sling. One of the crew went to push it to keep it clear of the hull-side but did not have enough strength. So, his teammate came to help but to no avail. Then the task leader added his weight to move the pump and the three of them pushed together. As the pump lifted it, caught on a protruding edge, the sling broke, the pump crashed to the sea dragging a cable which caused one of the men to fall overboard to his death and another to be seriously injured. The voiceover says: "Even seemingly small changes to a job can have a significant impact on the risks of a job."

This type of event is characterised by highly motivated workers, helping crew mates who are trying to complete a task. It is very easy to lose sight of escalating risks, when striving to get a job done. But noticing that the nature of the planned task has significantly changed should be taken as a warning sign that the risk level may also have changed. Could the second worker have paused before joining, or the task leader? While there are times when a task should be completely shut down or a procedure aborted, there are other situations where a pause may be all that is required to reassess the risks.

The idea of slowing down is gaining ground in other high-risk occupations. A Canadian surgeon, Carol-Anne Moulton has studied "slowing down when you should" during operations[3]. She found that more experienced surgeons were more likely to recognize the warning signs that the task needed to be slowed down to let them re-assess what they were doing. Here's a snippet of recorded conversation from surgeons[4] discussing an unexpected problem and pausing to think about the situation.

"Where's the hole? Can you see it? Well that's no good, if you can't see the whole hole, we can't repair it, can we? ... Right. Think. Let's just think here."

The warning signs suggesting that a pause is needed include an unexpected event, a plan not working as anticipated, confusion in the team, a task more difficult than normal or equipment not quite right.

It is notable that the prevailing mantra in NASA prior to the Columbia accident[5] was 'Faster, Better, Cheaper.'

I can't help but wonder if in hindsight they might now be thinking: *"if only we had paused for a minute..."*

2. https://www.ciaas.no/animations/
3. Moulton et al (2010) *J. Gastrointestinal Surgery,14*
4. Flin et al (2015) *Enhancing Surgical Performance.* p94
5. Starbuck & Farjoun (2005) *Organization at the Limit.* p30

Dr Rhona Flin is Professor of Industrial Psychology at Aberdeen Business School, Robert Gordon University and Emeritus Professor of Applied Psychology at University of Aberdeen. Her research examines human performance in high risk work settings (energy, healthcare and aviation) focusing on safety, organizational culture and non-technical skills.

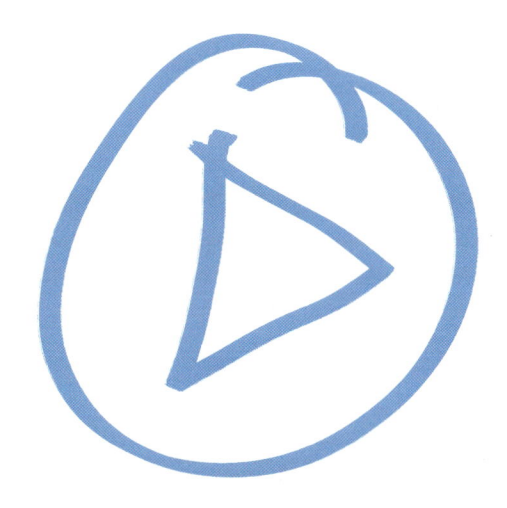

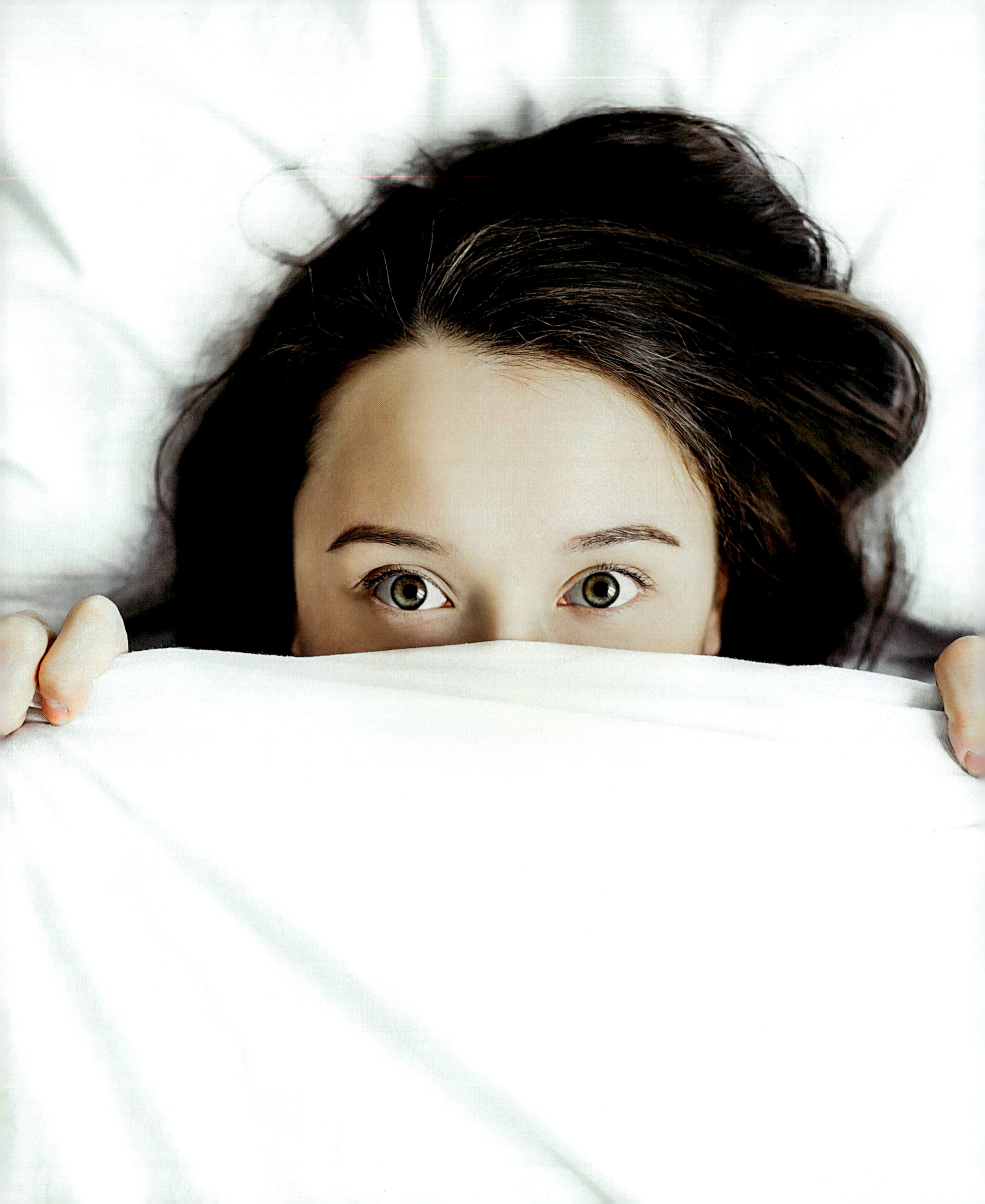

SLEEP BETTER AT NIGHT

GERALD FORLIN

I have been fortunate to work as a lawyer, lecturer and consultant in over 65 countries in many types of organizations and with different leaders.

I'm of the clear view that the vast majority of directors and senior officers want to do the right thing by safety and environmental issues. They want to do this for a myriad of reasons, including personal and corporate risk and reputation, moral obligation and simply doing the right thing by their workforce.

They often say to me that they "just want to be able to sleep at night."

They also tell me that they "just need to know how to do it!"

Experience tells me that one straightforward way of helping them sleep is to explain what the law is and using examples of real-life cases I've been in. Not just the high-profile catastrophic disaster cases, but also the ones that hardly make the press. I tell them the neutral facts and what happened to the various people in the litigation process, whether civil, criminal or an inquest.

In my career, I have acted for all types of litigants: large and small organizations, individuals, unions, families and the State. I try and explain how each party may perceive the process, result and aftermath. I try to give a 360-degree perspective. They often can grasp this quickly and easily, and they seem happier and enlightened by the process.

I then take them through the law and stress test their procedures and processes. I ask what they would feel like to be questioned on these three, maybe four years down the road. Would they feel confident and at ease? If not, why not? How could they change things now to be in a better place?

And I ask: "How can you sleep better at night?"

I also point out that what happens to their organization in one country or sector directly and indirectly effects the other parts of its operating base, in terms of legality and reputation.

I tell them about the various ISO standards and how regulators now base their investigations around the duties set out in them. They will need to be able to evidence that they have fulfilled the various requirements set out in the standards. There is no excuse or defence in not knowing something: the test is objectively what they ought to have known at the time, not what they said they knew.

This kind of exposure to the reality of litigation and enforcement leads to a change of attitude inside the organization and inside the people involved.

So, if you'd like to consider how an investigator or judge would view that omission to act or commission of that event, walk around your shop floor and see for yourself what's happening, and talk to people involved in all arenas. Leaders often feedback to me, sometimes years later, that this activity has greatly assisted them.

Every day do the right thing, potentially save a life and you'll sleep better at night!

Gerard Forlin QC was appointed as Queens Counsel in 2010. He has appeared in courts and worked internationally across industries including aviation, mining, chemicals, manufacturing, transport, finances and retail. Forlin has authored over 100 articles, chapters and books including *Corporate Liability: Work-related Deaths and Criminal Procedures*.

WE WILL
BE BACK!

BECOMING RESILIENT

RALF FRANKE

How can we make organizations and the world one percent safer? When I retrace ten years of Health & Safety evolution in my company, and examine the COVID-19 crisis, this is my answer: organizational resilience provides the key.

Organizational resilience is the proficiency to cope with dynamic, disruptive change, usually of a magnitude that can substantially impair, damage, or even destroy an organization. To be resilient requires a range of capabilities: an aptness to sense and anticipate changes in the ecosystem, a willingness to invest in preparing even for what seems less likely, and a readiness to quickly mobilize additional resources and reshuffle priorities when needed. And, perhaps most importantly: a continued focus on learning and adaptation.

These are the elements we have worked with in recent years:

- Ten years ago, we started out *laying the groundwork* – implementing the management system, structure, and processes.
- We *built a community of EHS people* endowed not merely with functional expertise but also with an open mindset and with management skills.
- We embedded *EHS Leadership in business* – an understanding that it is the CEO who is responsible for EHS and not the EHS Officer: so we make CEOs personally report fatalities to the Managing Board.
- We proceeded to connect Health Management with Occupational Safety, mainly in the field of mental well-being and psychosocial risk management. We collaborated with our HR development colleagues to integrate our health and safety topics into their concepts and programs.

Then COVID-19 entered the stage. The current crisis elucidates the importance of the resilience for an organization, and it shows where we stand regarding these capabilities. I am certain you've had similar experiences: in the context of leadership, we witness leaders committed to caring for their teams while others try to keep tabs on the behavior of their employees remotely, clearly not trusting them to do their best. In the context of work atmosphere, we experience hands-on, purpose-driven, agile collaboration across all hierarchies and organizational boundaries while in other scenarios, entire teams seem to be plunge into shock-induced paralysis and speechlessness.

With our program "Healthy and Safe@Siemens," we will focus on key aspects of organizational resilience: leadership, work climate, learning, processes and resources. The COVID-19 crisis is pushing us towards establishing resilience ever more firmly. We will accomplish that by focusing on our possibilities to exert influence on-site. Our program will support teams at sites to reflect on their experiences, to understand what worked out well, why certain actions do not show expected effects –to explore causal and contextual relationships as well as the consequences of what is being done.

To me, the decisive momentum of this evolution came about when we understood that for Environmental Protection, Health Management, and Safety, the highest potential for improvement is in tackling the system – not in optimizing its parts. And it was when we started to intensify our discourse with people in business as to how they actually do their work.

Dr Ralf Franke is Corporate Vice President and Corporate Medical Director, Head of Environmental Protection, Health Management and Safety, General Practitioner, Occupational Physician and Specialist for Occupational Safety, Siemens AG. Since 2009, Franke has been Head of Corporate Human Resources Health Management and Head of HR Environmental Protection, Health Management and Safety for Siemens.

PRUDENCE

FRANK FUREDI

Safety is an inherently subjective and diffuse condition, and perversely the very quest for safety often enhances the sensibility of insecurity. The mantra of "you can never be too safe" is a roundabout way of saying "be careful." Typically, the quest for safety turns every human experience into a potential safety issue. The exhortation "stay safe" has acquired a ritualistic quality that signals the message that we can never take our security for granted.

Experience shows that the quest for personal safety is not simply a response to external threats but increasingly a reaction to the internal turmoil associated with existential insecurity. Since existential insecurity is an integral element of the human condition in a world of uncertainty, it is likely that the quest for safety will be a never-ending feature of life. It's most important downside is that this quest distracts people from attempting to gain a measure of control over their affairs. Control requires a willingness to attempt to manage and live with uncertainty. And the most effective way of gaining that control is through the use of judgment.

The use of judgement is historically expressed through the virtue of Prudence, which refers to the ability to govern and discipline oneself through the use of reason. As a virtue, Prudence derives, in part, from the ancient Greek concept of *phronesis*, which translates into the term 'practical wisdom.' Prudence conveys an ability to recognize and follow the most suitable or sensible course of action.

On occasions, Prudence is misunderstood as a form of behaviour associated with the exercise of caution. However, Prudence is a multi-dimensional form of behaviour and possesses an ability to exercise sound judgement through a willingness to engage with uncertainty and take risks. So, unlike the risk-averse account of safety, Prudence is open to engaging with uncertainty and – when it judges it to be suitable – to take risks.

Though often portrayed as one of the four Cardinal Virtues, the ideal of Prudence can be understood as a quality of the mind that gains an understanding of the world through the use of reasoning and judgement. It is these characteristics that ensure that we do not simply react but also reflect on the nature of the threat they face. In this way, Prudence helps to provide a response that offers realistic balance between opportunity and threat. Most important of all, it provides society with hope and offers an antidote to the cultural power of fear. It recognizes that uncertainty can give rise to hope as well as to fear. And hope is arguably the precondition for making us feel secure, to the point that we need not always worry about our safety.

Dr Frank Furedi, author and social commentator, is emeritus professor of sociology at the University of Kent in Canterbury. His book *How Fear Works: The Culture of Fear in the 21st Century*, was published in 2018 and explores the distinct features of contemporary fear culture.

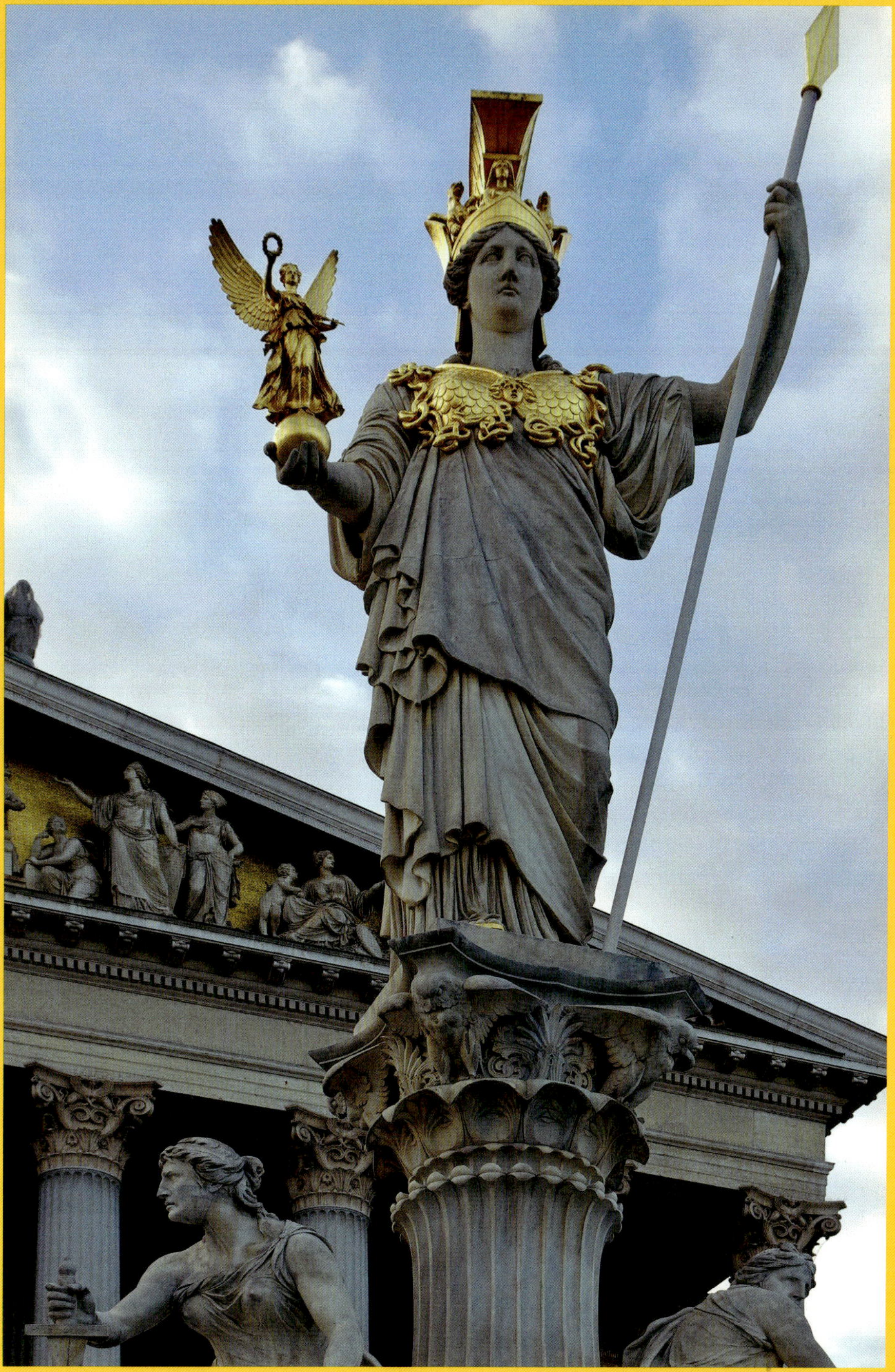

WHAT DO WE MEAN BY 'SAFER?'

RON GANTT

As I was backing up in my vehicle the other day, I noticed something interesting. My car has a back-up camera that turns on when you reverse. At the bottom of the screen, a message pops up when the camera turns on: "Search surroundings for safety."

It's an odd statement, really. Of course, the designers took it for granted that we would all know what they meant (*"Don't just stare at the screen, look around to make sure you don't hit anything you moron!"*).

But if we were looking for safety, what exactly would we look for?

It seems like a silly question, but what strikes me is that the safety profession does not have a straight answer. In fact, the best answer is that safety is "the absence of control of hazards or risks." Indeed, that is the idea that the designers of the backup camera had in mind – look for hazards and if there are none, then you must be safe.

This is an example of a belief in safety that is so fundamental that we take it for granted. The idea is that safety actually doesn't need to know anything about safety. All we have to do is identify *unsafety*, control or eliminate it, and then everything must be safe. Safety and unsafety are two very different things in this mindset. So why waste time understanding safety when all you need to do is eliminate unsafety?

But this mindset ignores the reality that hazards and risks don't exist in workplaces and society just to cause harm. Think about it: where do hazards and risks come from? They are the necessary side effects of work. Not because of some flaw in the system, but just because all work involves goal conflicts, uncertainty and scarce resources.

Yes, we can reduce or eliminate some of this, but our world is too complex to get rid of it all. What's left over are hazards and risks inherently related to the success of our organizations. We have to find a way to balance this complexity. Working in hazardous organizations is a problem that workers solve every day – and effectively!

This means we need to shift our thinking. We can't just focus on eliminating or controlling negatives if they are inherently tied to the good things in organizations. We need to understand how work gets done (different from how work *should* get done). Then we can start to identify ways to help workers solve the problems of work. By learning from what is happening in the organizations – and not just what goes wrong – we create the ability to ensure things go well.

This would go a long way in helping us identify exactly what we mean by "one percent safer."

The first step is expanding our focus beyond *unsafety*. Next, become curious about analyzing work more broadly, what it takes to get it done on a day-to-day basis, and consider the view "outside of the camera."

Ron Gantt is Director of Innovation & Operations at Reflect Consulting Group where he specializes in helping organizations improve performance and create resilience. He has almost 20 years of experience as a safety professional and degrees in occupational safety & health, psychology and safety engineering.

WHAT EXACTLY ↗
WOULD WE
LOOK FOR?

PRACTICE EMPATHY

HUMANISTIC BEHAVIORISM

E. SCOTT GELLER

Behaviorists manage behavior by applying positive consequences for desirable behavior and removing positive consequences for undesirable behavior. In contrast, humanists target people's intentions, and focus on discovering personal perceptions, motives, and self-concept. The humanist's clinical approach is nondirective, with the therapist doing more listening than directing. Behavioral therapists are directive while defining behavioral consequences that can be changed to increase desirable behavior and decrease undesirable behavior. You can optimize the injury-prevention impact of behavior-based safety (BBS) by applying select principles from humanism. Let's consider three of these principles.

Practice Empathy
The critical behavioral checklist (CBC) used in BBS to observe safe and at-risk behavior is not handed down by OSHA, management or a safety professional. No, a team of workers on a particular job develops the CBC and modifies it whenever their job and/or the environmental conditions change. Moreover, when coworkers, safety professionals and supervisors give an employee corrective feedback for observed at-risk behavior, they don't deliver behavior-change directives. They ask questions to understand the rationale for the at-risk behavior, aiming to discover features of the situation that could be altered to facilitate occurrences of the safe alternative. This is empathy in action.

Facilitate Self-Accountability
Rather than telling an employee what safe behavior should replace an observed at-risk behavior, humanistic observers ask the worker what s/he could do to reduce the probability of an injury. Similarly, after identifying a work problem, humanistic supervisors do not specify a resolution. Instead, they challenge the relevant employees to discuss possible solutions and propose an action plan. This strategy for facilitating self-accountability reflects humanistic self-determination theory. Specifically, perceptions of autonomy (or choice), competence, and relatedness (or interdependence) enhance self-accountability.

Appreciate Maslow's Hierarchy of Needs
According to the hierarchy of needs created by the humanist Abraham Maslow, we are first motivated to fulfill our physiological needs—basic survival requirements for food, water, shelter, and sleep. After these needs are under control, we are motivated to feel secure and safe from potential dangers. Next, we have our social-acceptance needs—to have friends and feel a sense of belonging. Then our concern focuses on self-esteem—to earn self-respect and feel worthwhile.

After enjoying a perception of self-esteem, we become self-actualized—we realize our full potential. While many believe self-actualization is atop Maslow's need hierarchy, it is not. Near the end of his life, Maslow placed another ultimate achievement at the top—self-transcendence. We are the best we can be when we reach beyond our own self-interests and contribute to the needs of others. Isn't this what safety leaders do on a daily basis? They intervene whenever possible to keep others safe from personal injury.

How satisfying to realize that you reach the top of Maslow's Hierarchy of Needs whenever you act on behalf of another person's safety. When reaching this highest need level, we satisfy our lower-level needs that never get completely satiated—social acceptance, self-esteem and self-actualization.

Dr E. Scott Geller, Alumni Distinguished Professor, just completed his 50th year as a teacher and researcher in the Department of Psychology at Virginia Tech and Director of the Center for Applied Behavior Systems. He a co-founder and Senior Partner of Safety Performance Solutions, Inc., and GellerAC4P, Inc.

SEE AND BE SEEN

FRANCOIS GERMAIN

When a business unit needs a culture change to improve safety performance, there is no magic answer, but there is a prerequisite: it must start from top management.

Below is a summary of the strategy that we used in 2018 when a large business unit of total needed to respond to a degradation in safety performance. The strategy – defined with all plant managers – was simple, and built on 4 key pillars:

- **Visible leadership**: reinforce competency in safety leadership
- **Visible commitment**: field presence, feedback and intervention
- **Visible recognition**: recognition rituals in each factory
- **Visible standards**: expect high level of requirements on lifesaving rules: work at height, traffic, provision and work on energy (those causing accidents in the BU)

It was then rolled through all the sites with a rapid implementation.

The safety leadership training session was given to all leaders, the plant management team, together, were then expected to go on the field to engage safety dialogue with employees and contractors on a weekly basis (even more frequent for smaller sites). The other managers embarked for the similar ritual. The strategy worked well, the results followed: accident rate divided by 3 in less than 2 years (Total Injury Rate <1)

The success factors were:

- 'Zero is possible' mindset: *it's our choice and we are convinced that we can do it*
- Move from compliance to conviction: *I want to work safety versus I must comply*
- Visibility of the line management caring about employees and involved in Health and Safety aspects
- Great emphasis given to field presence: *we convince by our presence*
- Coach versus cop attitude on the field: *less audit, more dialogue*
- Count and tell more when we were good rather than only failure or lucky event
- Be rigorous on application of standards: use of check list for all employees to know what to observe at a minimum
- Clear commitment from everyone: commitments were publicly posted and signed by individuals

The business unit succeeded in their culture change, which enabled in parallel improvement in Operational Excellence culture e.g. LEAN manufacturing.

It is all about leadership and passion for caring for our employees and contractors.

Francois Germain holds a Masters' degree in General Engineering. His career spans power generation, chemicals, refining and offshore. In 2014 he joined Total Refining and Chemicals as VP Health & Safety where he leads global Safety Culture transformation programs for several Business Unit's within Total.

GIVE PEOPLE CONTROL

ISAAC GETZ

British psychologists Bosma, Stansfeld and Marmot studied stress levels of more than 10,000 British civil servants. Their findings showed that men perceiving little control over their jobs are 50% more likely to develop heart disease than those who feel in control. For women, the risk is 100% higher.

Lack of control over tasks triggers employees' emotional reactions, such as anger or anxiety. Sometimes reactions are constructive, such as negotiating a better work schedule. Most often they are destructive: flight or fight, accompanied by increased adrenaline, blood pressure and heart rate. While helpful throughout human history to cope with predators, these companions aren't useful in fleeing the job or fighting the boss. The longer stressors remain 'unresolved,' the more their damage, called 'psycho-somatisation' leads to stomach disorders, back pain, musculoskeletal problems, headaches, loss of sleep and energy, and emotional distress. In the long term, they lead to heart and vascular diseases, and in the worst case, death.

The costs from stress-induced absenteeism have been estimated for US businesses as between $150 billion to $300 billion per year, with UK costs proportionally comparable. Presenteeism extols an even bigger cost than absenteeism. Though ill, employees come to work but are 50% less productive, prone to mistakes and chronically tired.

The root cause of all this suffering is the unmet universal human need for responsibility and freedom – 'self-direction' – in work. In traditional hierarchic bureaucracies only people with control can potentially escape from supervisors or procedures; most employees at the bottom have their need for control over their tasks denied.

There is a way these hidden costs can be eliminated for good. *Give people real control over their work.* Stop telling them how to do their work, and the stress levels will go down. Absenteeism will go down. Hidden stress-related costs will go down. Engagement will go up.

Of course, this is hard to accomplish in command-and-control companies. A way leaders are achieving this is through corporate liberation[1]. A liberated company is one in which employees are free and responsible to take actions that they—not their bosses, not procedures—decide are the best for the company[2]. The two key ingredients of performance – employee initiative and potential – which are traditionally stifled, are freed up in liberated companies. Hundreds of companies and public service organizations have undergone these transformations recently, with tremendous human and, consequently, economic results.[3]

Since they are willing to get up in the morning and go to work to give their best, liberated company employees are outperforming their old-style competition. Put differently, freeing a company's people to act not only eliminates the burden of hidden stress-related sickness costs and deaths, it also dramatically improves wealth of both companies and nations.

1. Brian M. Carney & Isaac Getz, *Freedom, Inc.*, Crown Business/Random House, 2009, revised and expanded edition, 2016, 400 p. (ISBN 978-2-08-138021-9)
2. Isaac Getz (Summer 2009), « Liberating Leadership: How the Initiative-Freeing Radical Organizational Form Has Been Successfully Adopted, California Management Review, Vol 51, N°4, pp. 32-58 », on cmr.ucpress.edu (consulted on May 10, 2017)
3. "Adam Gale, "Employee engagement: The French way", Management Today, January 24, 2017". www.managementtoday.co.uk.

Dr Isaac Getz is a Professor of Leadership and Innovation at ESCP Business School. He is an internationally acclaimed management expert, active public speaker and author of *Freedom, Inc., Leadership Without Ego* and *The Altruistic Corporation*. He was a Visiting Professor at Cornell University, Stanford University and the University of Massachusetts.

DON'T 'PASS THE BUCK' ON OCCUPATIONAL HEALTH

ALISTAIR GIBB

For many decades in many countries across the world, we have whispered 'health' and shouted 'SAFETY.'

Despite using the phrase Occupational Safety and Health (OSH), the emphasis has been on the more immediate 'safety' concern. However, even in high-hazard industries like construction, many more workers die from occupational health hazards than safety hazards. And this situation continues in countries that have focussed on and made significant improvements to their safety performance.

In the UK 30 years ago, construction managers focused on time, cost and quality, with safety officers acting as police officers trying to force managers to address hazards. If a manager was asked about safety they would 'other' the problem, or 'pass the buck' to the safety officer, claiming "It's not my problem – just tell me what rule I must obey!"

In recent years, in countries including the UK, we have matured such that construction managers have started to accept that they have the responsibility to plan, organize and manage the works in a manner that facilitates safe behaviour. Safety is now seen as 'part of the job' for managers and the safety professional is now an advisor, called upon to offer expert advice when needed.

However, when it comes to occupational health, and particularly mental health, managers still 'pass the buck' to the health and safety advisors. To make things worse, because of their limited knowledge and experience of health issues, these advisors pass things on again to health experts such as occupational hygienists.

We need to up-skill managers to understand enough about occupational health and challenge the predominant culture such that they accept the holistic view that they have a key role in influencing workers' health and wellbeing *as well as their safety*. And we need to train and develop health and safety advisors to be able to provide the support and encouragement to achieve the same degree of improvement in health and wellbeing that we have achieved in reducing accidents.

So, business leaders, don't pass the buck, ensure that your managers 'own' occupational health – as well as safety – and treat it as "the way we do things around here".

Dr Alistair Gibb is Professor of Construction Engineering Management, and Researcher and Research Manager in construction innovation and Occupational Safety and Health (OSH) at Loughborough University. A Chartered Engineer and Chartered Construction Manager, Gibb joined Loughborough University in 1993 following a career in civil engineering and construction management.

DON'T PASS
THE BUCK

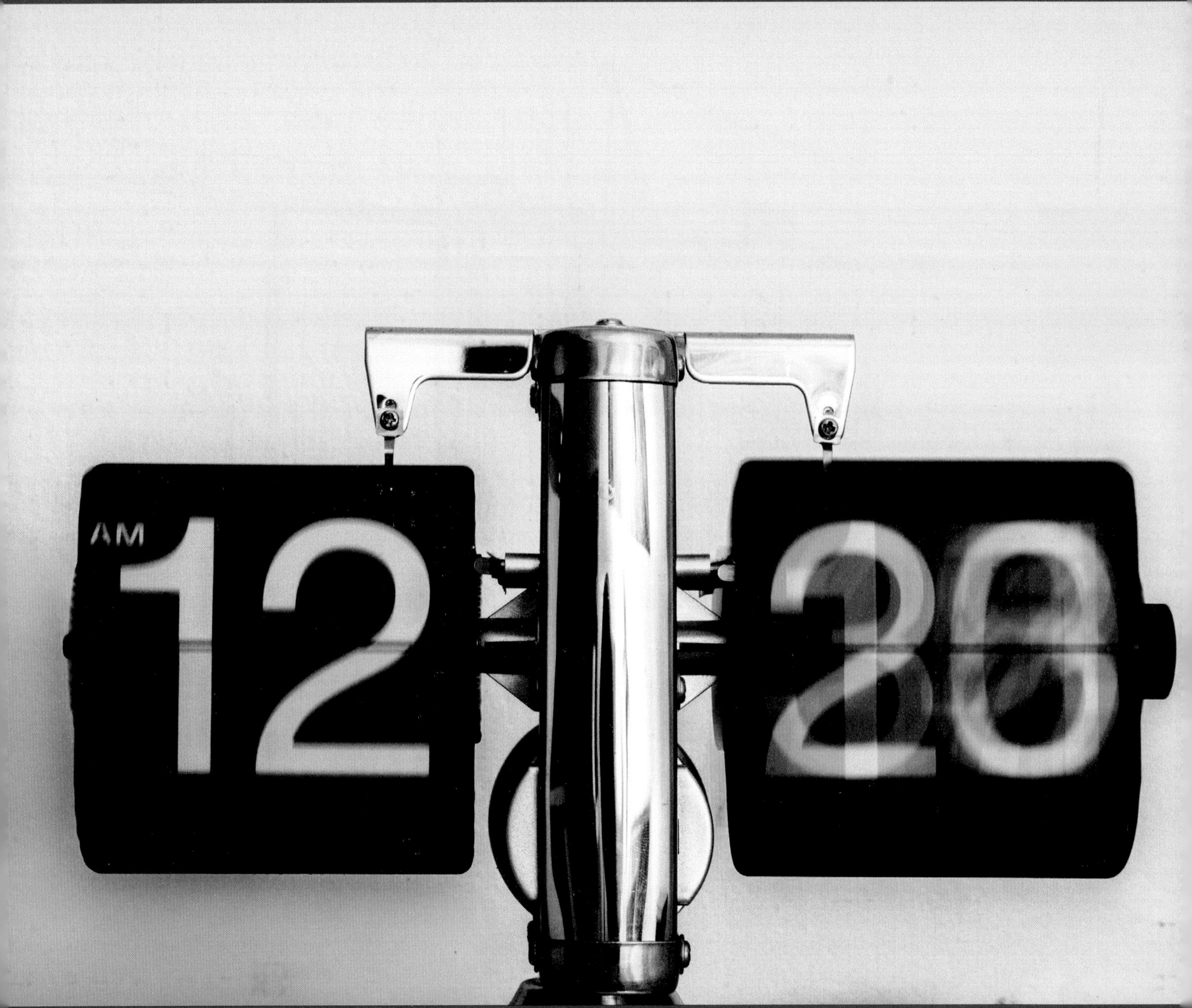
AM
12

THE LEADER'S TOOLBOX

GERD GIGERENZER

Senior managers routinely need to make decisions or delegate decisions in an instant, after brief consultation and under high uncertainty. For their work, classical decision theory is of limited help. In the words of Henry Mintzberg[1]:

"Managers work at an unrelenting pace; their activities are typically characterised by brevity, variety, fragmentation and discontinuity; and they are strongly oriented to action."

The rules of thumb the top executives rely on are often unconscious satisfying the definition of intuition. Inspired by my findings on rules of thumb in the book *Gut Feelings*, the former president of Florida International University, Modesto Maidique, developed an innovative view on the nature of leadership. Every senior manager comes to the job with a personal 'adaptive toolbox.' This toolbox contains a set of rules of thumb derived from personal experiences and values. They are the basis for making decisions about persons, strategies, and investments in a world that puts a premium on efficient use of time.

Good leadership consists of a toolbox full of rules of thumb and intuitive ability to quickly see which rule is appropriate in which context. Here are six rules of thumb that are the result of an interview with Ray Strata, Chairman of Analog Devices, in September 2010. He took personal risks to move his company into a new field. The rules are different for dealing with people and business strategy.

People

- First listen, then speak
- If a person is not honest and trustworthy, the rest doesn't matter
- Encourage and empower people to make decisions and take ownership

Strategy

- Innovation drives success
- Analysis will not reduce uncertainty
- When judging your plan, put as much stock in the people as in the plan

These rules are mostly intuitive, meaning that the leader cannot easily explain them – just as we speak a native language without a second thought, but often flounder when asked about grammatical details. True leadership means intuitively understanding what rules work in what situation.

1. Mintzberg, H. (2009). *Managing. San Francisco: Berrett-Koehler.*

Prof Dr Gerd Gigerenzer is a German psychologist studying bounded rationality and heuristics in decision-making. Gigerenzer is Director Emeritus of the Center for Adaptive Behaviour & Cognition (ABC) at the Max Planck Institute for Human Development in Berlin and Director of the Harding Center for Risk Literacy at the University of Potsdam.

CARLOS'S BRAIN: EMPATHY AND NEUROSCIENCE

CHRISTOPHE GILLET

In the back of a packed and enthusiastic room, I listened to the expert in neuropsychiatry and quantum physics explaining brain sections, with a full colour scan for support: "As you can see on this scanner, power contributes to reducing the areas of empathy and compassion in the brain."

Through this individual, science had just spoken … Indeed, we could clearly see on the picture that the suggested area of empathy was in fact being reduced.

Well, gosh, what a surprise! I'm speechless...

So one question runs right through me:
Science has just offered the evidence that power can reduce empathy. (Assuming the expert didn't get the wrong area, or color, or image, or even brain.) And yet: *So what?* What can we do with this information?

We know that Carlos' brain (a random name, but it could be mine or yours) is going to be impacted by his supreme functions… So what can you do?

I'm curious. I'm really curious.

Making the world one percent safer implies among other things, *a strong human focus*, relying upon empathy and compassion. The paradox is that it seems that the rat race of the workplace most of us play makes us less capable of going in this direction, especially those of us with power, or in leadership positions.

As a leader with the knowledge that power reduces empathy and compassion, how can you reframe your own mindset to make your workplace and workers safer? How can a reframed mindset encourage the empathy required to put yourself in another's shoes? Simply, now that you know you share the characteristics of Carlos' brain:

What are you going to do about it?

Christophe Gillet operates in the fields of Innovation, Transformation & Uncertainty focusing on human mental patterns and organizational aspects. In a former life, he was Director of Innovation for SONY Business Europe. He contributes to learning organizations globally including INSEAD, IMD, CEDEP, University of Cape Town and Duke CE.

FIRE! 4 VITAL MINUTES

DAVID GOLD

The past ten years are marked by a number of large life-loss fires in both the developing and developed world, from factory fires in Asia in the garment industry to residential building fires such as Grenfell Tower fire in London. The loss of life, the amount of physical and psychological injury and the financial losses are considerable, and in some cases, devastating.

Yet fire safety is rarely discussed in an organization's C-Suite, as it takes on a much lower priority than issues such as finance or productivity. It is often looked upon as a necessary expense, not an asset. We know there are ethical, legal and financial reasons to address fire safety, however, there are often companies with a pervasive and complacent attitude that says: "it is most unlikely that this will happen in my organization".

Although we must be diligent in focusing on fire prevention, across industries, fire *evacuation* is a critical first step. In an audit of your organization's fire safety procedures, the first and most important box that must be ticked is:

When the evacuation alarm is activated (automatically or manually), does everyone get out of the building and into an area of safety (a fire assembly point) within four minutes?

In buildings everywhere, from the informal sector (such as in a developing country) to a building that is part of a multinational enterprise, the growth of fire knows no economic, social or physical boundaries. Fire research clearly demonstrates that after four minutes from the when the fire starts (ignition), the ability to leave the building to an area of safety is increasingly compromised by increases in heat, smoke, flames and fire gases such as carbon monoxide, as well as a decreasing amount of breathable oxygen in the air. There is no time for those being evacuated to be complacent (e.g. confirming the alarm, gathering belongings or searching for friends) if they are to be able to get out safely.

So, as an organizational leader, what three steps can you initiate to tick (and sustain the tick in) this first box?

Ensure that:

1. There is a functioning evacuation alarm that allows everyone to know they must evacuate and through clear, unobstructed, well-marked, well-lighted primary and secondary escape routes leading to their designated fire assembly point.
2. Workers, visitors, contractors and others are initially and continually made aware of fire emergency procedures and their importance in a language that they understand and are knowledgeable of their primary and secondary escape routes and their designated fire assembly point.
3. Fire evacuation drills with total participation are carried out and evaluated twice a year (one of which continues all the way to the designated fire assembly area).

Organizational leaders must be prepared to address fire safety issues by fully integrating fire safety into the mindset of the organization. Ensuring the above three points and bearing in mind the four-minute time frame is a solid first step.

Dr David Gold, CFIOSH, MIFireE, is an international consultant in occupational safety, health and fire and a Vice President of the Institution of Occupational Safety & Health. He served as a Senior International Official for the ILO and has authored numerous articles on fire safety, psychosocial issues and diving safety.

IT'S ALL ABOUT THEM

MARSHALL GOLDSMITH

For years people have been asking me, "What makes a great leader?"

Well, I don't believe that extraordinary intelligence is an attribute that is shared by all great leaders – not least because 'super-smart' leaders can be impatient and judgmental, which hinders their ability to get the best out of their teams. Sure, leaders do need a degree of intelligence, and that will vary on the business sector the leader finds themselves in. If there's a significant difference between the IQ of a leader and the people around them, then that's where things get tough.

I'm reminded here of one of my own personal favourite leadership lessons, which comes from Peter Drucker. Peter taught me that the leadership mission is not to prove how smart we are, or how right we are, rather our mission is to make a positive difference.

Here's the first step: to be a great leader, you'll need to understand that it's not all about you: it's all about them. You have to let the people be the heroes.

So how do you turn regular folks into heroes?

It's not that difficult, really.

Start by getting input from the people around you, close your mouth and listen, then recognize the contributions of others, and thank people for their efforts. Want to improve workplace safety? Ask your people what they'd like to see change. Ask them where the biggest risks of them getting hurt are, and then ask for their suggestions on how to stay safe.

Next, know that great leaders don't worry too much about what he or she looks like. Instead they worry about helping the people around them to look great, feel great, and be great. Remember, it's all about them.

Finally, great leaders have a strong sense of self-awareness. Over the last 50 years I've come to learn that with every problem I've ever had, I just had to look to one place to see the source of the problem. I looked in the mirror. What happens when you look there too?

Dr Marshall Goldsmith is a member of the Thinkers 50 Hall of Fame and is the only two-time Thinkers 50 #1 Leadership Thinker in the world. He's been ranked as the World's #1 Executive Coach and was the inaugural winner of the Lifetime Award for Leadership by the Harvard Institute of Coaching. His books have sold over 2.5 million copies worldwide.

SMITH STREET

THE QUESTION, THE EXPRESSION AND THE PICTURE

SONNI GOPAL

Addressing and improving risk and safety performance within an organization requires a helicopter view, a deeper dive and incremental improvement. Three things that require three very different, though interconnected, actions.

The Question

It was Desmond Tutu who once wisely said that *"there is only one way to eat an elephant: a bite at a time."*

This guidance on the importance of pace resonates today. To improve risk and safety performance, it is important to step back and keep asking the question, "Are we taking a bite at a time?" Focused, small incremental efforts have been proven to yield sustainable safety performance improvement.

The Expression

In 1956, Charles R. Schwartz authored the article 'The Return-on-Investment Concept as a Tool for Decision Making'. In this article, he stated:
"We will do less guessing; avoid the danger of becoming extinct by instinct; and, by the adoption of one uniform evaluation guide, escape succumbing to paralysis by analysis."

Risk assessment is defined as "the process of risk identification, risk analysis and risk evaluation." (Ref: ISO 31000 Risk Management Guidelines) and it is quite common to over-analyze, or over-think, during risk assessments. Essentially, paralysis by analysis is the state of over-thinking a situation to such an extent that a decision or action is never taken.

I use the expression 'paralysis by analysis' when conversations start to drift aimlessly. After just a few mentions, the expression becomes an anchor point. Nearly everyone in the risk assessment team then starts to adopt the expression. It is a great 'pause for thought' catalyst.

The Picture: 'Safety Continuum'

Whatever the nature of your organization, there are essentially four primary activity areas, and these are:

- Design
- Engineering
- Procedures
- People

Some organizations may not do any Design nor Engineering. However, they will have People and this in turn will drive the need for Procedures.

For your organization, take into consideration this picture and then take a helicopter view. After deciding which aspects are applicable, you can proceed with a deeper dive for granularity and by taking "a bite at a time."

During the deeper dive review, it is inevitable there will be potential for over- analyzing – and this is when we recall and reinforce our expression 'paralysis by analysis'. Forming this habit loop for the organization with the *Question*, *Expression* and *Picture* will contribute significantly towards risk and safety performance improvement.

Sonni Gopal has over 30 years of international, multi-sector experience in Risk Management, Workplace and Process Safety. A global digital networker, he uses digital and social media platforms to openly share knowledge and experience with risk and safety management practitioners.

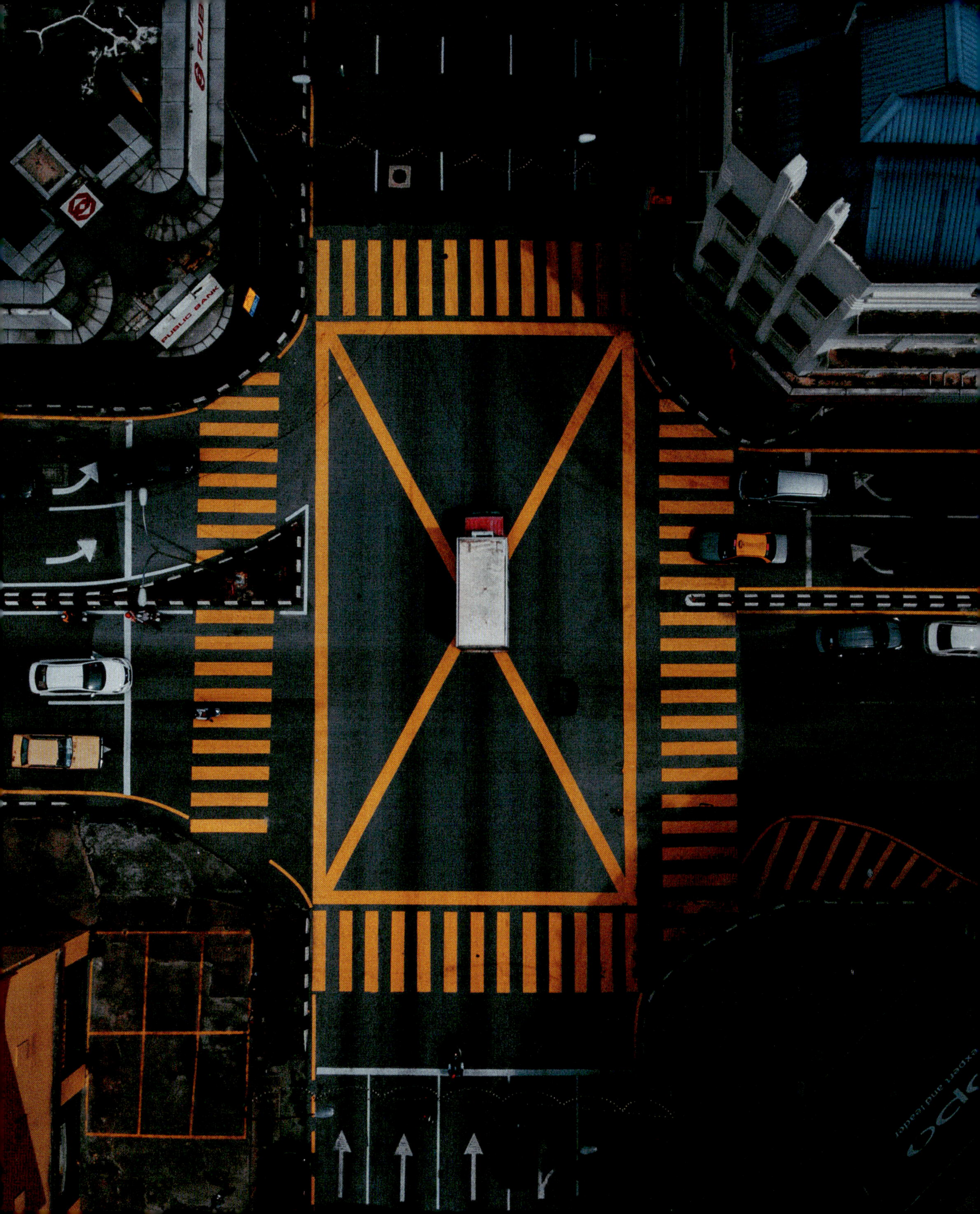
PUBLIC BANK

FOR WANT OF A NAIL THE SHOE WAS LOST. FOR WANT OF A SHOE THE HORSE WAS LOST. FOR WANT OF A HORSE THE RIDER WAS LOST. FOR WANT OF A RIDER THE MESSAGE WAS LOST. FOR WANT OF A MESSAGE THE BATTLE WAS LOST. FOR WANT OF A BATTLE THE KINGDOM WAS LOST. AND ALL FOR THE WANT OF A HORSESHOE NAIL.

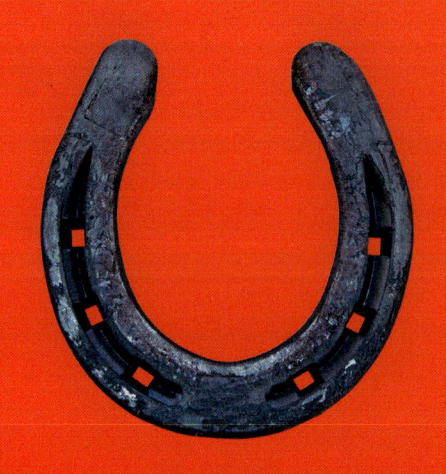

PREVENTING THE LOSS OF KINGDOMS

JOP GROENEWEG

In the early days, the stable boy or the blacksmith who made the 'human error' that led to the missing nail would have been fired.

After it was acknowledged that this wasn't very effective, the focus shifted to the availability and compliance with procedures, training and awareness about the importance of nails.

As kingdoms kept falling, it became apparent that it is important to find out 'why' the nail was missing. The organizational factors leading to the lack of nails and shoes became the next target, e.g. ensuring that the balance between the advantages of keeping supplies and the burden of the associated costs didn't shift to the latter. Systems were put in place to control the risks involved in fighting wars. Kings publicly proclaimed that winning battles was their first priority. A substantial reduction in fallen kingdoms was the result, but the road ahead was unclear.

The nursery rhyme raises an important question that could provide guidance: *How is it possible that the loss of a single nail can have such devastating consequences? What are the causes of such an unacceptable vulnerability?* Kingdoms are complex organizations and it is fundamentally impossible to prepare for every single scenario that could cause it to fail. Those involved in the battle never saw the fall of the kingdom coming: how could they possibly have foreseen that a nail or a shoe could have such terrible consequences? Kingdoms should therefore be organized in such a way that they are able to cope with the loss of nails, horses and riders. If the focus stays limited to factors leading to disastrous losses, the factors will be missed that raise red flags about the all-important state of the kingdom itself. Everybody, from kings, traders, nuns, farmers and cooks to knights and sailors should speak up about whatever they consider a deviation of what can be expected in a well-run kingdom.

In this way, curiosity, creativity and open-mindedness are stimulated, so daring suggestions to further improve the way the kingdom is run are shared without fear of criticism, even if some of the initiatives will inevitably be less successful than expected. Every action becomes an opportunity to gain new insights that are shared with the world. By amplifying each other's personal, social and professional capabilities, the result will be just and prosperous kingdoms where catastrophic failures become a distant memory.

The key to success is the creation of organizations where the need to excel at every activity is self-evident: effective, efficient operations and therefore excellent safety performance will follow.

Dr Jop Groeneweg is Professor of 'Safety in Healthcare' at Delft University of Technology in the Netherlands. He's a human factor specialist at Leiden University and senior researcher at Dutch research institute TNO. His current research focusses on increasing psychological safety in teams to stimulate organizational learning, improving operational and safety performance.

BE A PARADOX-SAVVY LEADER

GUDELA GROTE

Core to safety is the ubiquitous tension arising from concurrent demands for stability and flexibility in the face of external and internal uncertainties. Centralized decision-making, hierarchical control and adherence to predefined procedures promise stability. Decentralized decision-making, competent local actors and improvisation in unprecedented situations promise flexibility. In order to reconcile this tension between safety and autonomy, leaders need to be clear about a number of important distinctions.

Process versus personal safety

Are the risks to be managed inherent to the primary work task (like keeping an aircraft in the air) and/or are they threats to workers' health?

Intrinsic versus extrinsic motivation

Are work tasks sufficiently interesting and meaningful to instigate motivation of themselves or are additional external rewards and controls needed?

Safety compliance versus safety participation

Do work tasks only require compliance with rules and procedures or is proactive initiative needed?

Operational autonomy versus higher-order autonomy

Do work tasks require local decision-making and/or can workers be involved in determining general rules to follow?

Leaders need to adapt their behavior depending on which combinations of these factors are relevant at any point in time. This is what paradox-savvy leadership is all about. Rather than establishing one particular leadership style, leaders have to develop a repertoire of behaviors in order to adequately address varying situational demands, and in particular, different degrees of uncertainty. Such leadership behaviors have to range from fostering stability and control through personal direction to delegating responsibility and giving up control when high flexibility is required. This also entails being sufficiently open-minded and perceptive to capture contradictory demands arising from the fundamental tension between safety and autonomy.

So, here are five guiding principles for becoming a paradox-savvy leader:

Principle 1. Transformational leadership – that is motivating employees through inspiration and charisma – gives workers a sense of purpose when personal safety is at stake, and the work tasks in themselves cannot create sufficient safety motivation.

Principle 2. Shared leadership by dynamically delegating operational autonomy to workers allows them to manage highly uncertain situations where process safety is at stake, workers are intrinsically motivated, and a high degree of safety participation is needed.

Principle 3. Continuous sensemaking by leaders helps workers to understand concurrent demands on safety compliance and safety participation and eases transitions between more centralized and decentralized modes of operation.

Principle 4. Participatory leadership that encourages workers' involvement in rule making promotes process safety by bridging rote rule following and proactive rule adaptation and increases intrinsic motivation for personal safety.

Principle 5. Culture-sensitive leadership builds on a shared commitment to safety to increase intrinsic motivation for personal safety and on shared cultural norms rather than rulebooks to set bounds to worker empowerment for process safety.

Dr Gudela Grote is Professor of Work and Organizational Psychology at the ETH Zürich, Switzerland. She received her PhD from Georgia Institute of Technology. Her research focuses on increasing flexibility and virtuality of work in the context of leadership and coordination in high-risk and innovation teams and socio-technical system design.

THE ART OF CULTURE

FRANK GULDENMUND

The thing with culture is, well, that it is not a thing. It is a label people put on groups of people having similar (and this can be a very long list) rituals, symbols, jargon, clothing, hair, *whatever*. What's important here is that all these things people share are not *because* they have, what we label, "a similar culture." Culture is not a cause; it cannot be. It is a label.

I know this is confusing, it has confused me for years. But I think I can explain this much better now for myself and – hopefully – for others, too.

We are all familiar with labels. We put them on just about everything. Everybody does it, therefore it must be innate and crucial for our survival. A particular form of labeling is profiling. We can imagine that this is important for us, because of these profiles we can avoid people we think might hurt us. Labelling is important too, as we can classify things into 'familiar' or 'strange' or 'pleasant' and on and on. It is easy and does not require much effort as we have many labels and profiles readily available.

I think most of us have heard of the label 'Impressionism'. Impressionism is a form of art that was particularly popular at the end of the 19th and the beginning of the 20th century. I'm not sure what kind of images you will now see with your mind's eye, but I see sunny French landscapes or small villages, kind of blurred, painted by men with beards and hats on. There are trees and shadows and blue skies, and so on. This is Impressionism, and Musée d'Orsay in Paris is full of it. However, and this is important, these painters did not paint the way they did *because of* Impressionism. No, Impressionism is a label we pin onto this kind of art.

Can we measure culture?

No, we can't.

Can we measure 'Impressionism'?

No, we can't. We can measure the canvasses Impressionists used, determine their palette, we can calculate their ages, determine the circumference of their heads, shoe sizes, and so forth. But what does this tell us about Impressionism? Not much, I'm afraid. Can we appreciate Impressionism? Yes, we can. Can we recognize Impressionism? Yes, that too (there is indeed a funny experiment where researchers taught pigeons to distinguish between Impressionist paintings and Cubist paintings).

Can we study culture?

Yes, we can. We can think about what makes a culture *this* culture. We can draw borders between what belongs to this culture and what does not (e.g. Picasso is not an Impressionist). We can describe a culture to the best of our abilities. We might dislike particular aspects of a culture as much as we dislike particular paintings, or painters. But this is a judgement, not a fact.

Next time you visit Paris, do go to the Musée d'Orsay and appreciate culture anew.

Dr Frank Guldenmund is a psychologist that has been studying and writing about organizational (safety) culture for many years now. He works at the Safety Science & Security Group at Delft University of Technology in the Netherlands.

CAN WE MEASURE CULTURE?

CARING IN CRISIS

JUDITH HACKITT

I am writing this on the day that manufacturing and construction workers in England are being encouraged to return to work after several weeks of pandemic lockdown. The question of what we can do to make things safer for us all feels very pertinent. In fact, I had the chance to share my thoughts on Radio 4's 'Today' programme.

For me, the one thing we can do is simple and clear: We all need to look out for and care for one another.

Being safe has never been about doing as we are told. If we simply do what the rules tell us to do, it means we are not 'bought in' – someone else has decided what is or isn't right, and we're doing their bidding – even if we think it's wrong or we've got a better idea. Even if the rules are right, we will forget them sooner or later because we're not bought in.

COVID-19 has reminded us all of just how highly we value those who care for us. But caring is not confined to hospitals, and it's not confined to being treated when we are unwell or injured. The best caring is the small things we all do every day to value each other and to remind us of our worth, not just as employees but as colleagues, as friends.

If there's one thing I've missed more than anything in the last few weeks, it's been the face-to-face contact with the people I work with – because they are people I like and who I value as friends. Zoom and Skype are great, but it's not the same. But have you also noticed how many messages we receive currently open by asking how we are and close by reminding us to "Stay safe"?

If we want the world of work to be safer, we need to continue to show that level of caring and go even further. "How are you?" needs to be followed up with "Is there anything I can do to help or support you?" Or perhaps: "What's on your mind? Can I help?" When I go back to working with people face-to-face, I am going to say to people "I've missed you, and it's so good to see you in person again".

Do I really think this will make a difference to safety? Yes I do, because if we all look out for and look after the people we work with, we will create a caring and safe culture.

Dame Judith Hackitt DBE FREng has a passion for safety which is recognized around the world. Her career has included chemicals manufacturing and almost a decade as Chair of the British workplace health and safety regulator. She now holds a number of non-Exec roles in Manufacturing and Engineering.

LOVE IS A human RIGHT
Support LOVE
LOVE wins
LOVE is LOVE
live and let love
LOVE IS A human RIGHT
Support LOVE
Support LOVE
LOVE wins
LOVE is LOVE
live and let love
LOVE IS A human RIGHT
LOVE IS A human RIGHT
Support LOVE
LOVE

FLEXIBILITY RATHER THAN RULE

ANDREW HALE

This flowchart summarizes my views on how organizations should view their safety rules and procedures; not as things carved in stone to be enforced willy-nilly, but as bases for learning and change. To be a true learning organization demands constant review and improvement.

Boxes 1, 2 and 4 in the diagram should therefore drive the process of rule management, which should run constantly and should guide and motivate involvement of all line and staff employees, process designers and contractors.

Reporting of incidents, accidents and other significant deviations are to be encouraged rather than hidden. The more they are reported, the safer the workplace will be, research has shown. Any investigations must be conducted, not as a search for the guilty, but as an opportunity to improve for the fallible. 'Toolbox talks', envisaged as top-down instruction and motivation are to be turned into 'toolbox listens' where front-line supervisors, eager to unlock and harness the knowledge of operators, hear what real-life is like, with its many exceptions.

Adapted from Hale, A.R. & Borys, D. 2013. *Working to rule or working safely? Part 2: The management of safety rules and procedures*. Safety Science. v55, pp222-231

Professor Andrew Hale has researched and taught the theory and practice of safety science his whole working life, first in the UK and then the Netherlands, latterly as Professor of Safety Science at the Delft University of Technology. Hale was also chief editor of *Safety Science* from 1993-2008.

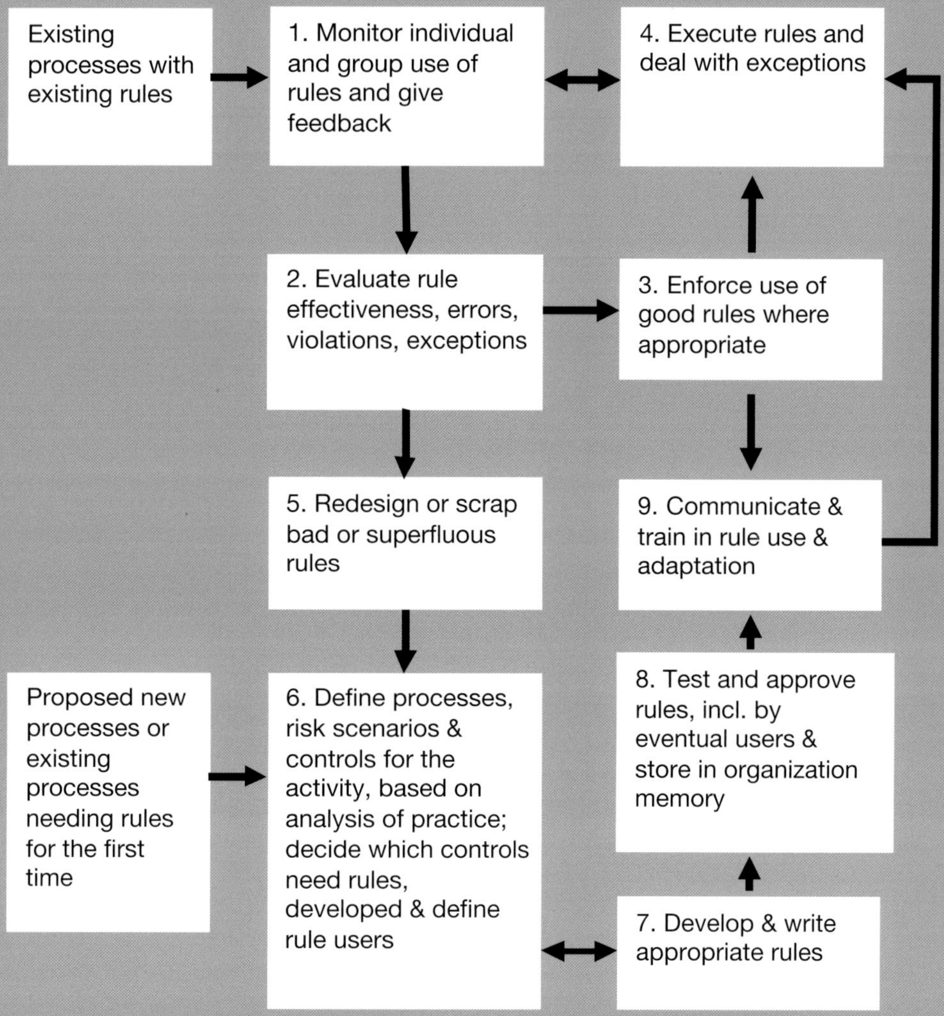
Existing processes with existing rules
1. Monitor individual and group use of rules and give feedback
4. Execute rules and deal with exceptions
2. Evaluate rule effectiveness, errors, violations, exceptions
3. Enforce use of good rules where appropriate
5. Redesign or scrap bad or superfluous rules
9. Communicate & train in rule use & adaptation
Proposed new processes or existing processes needing rules for the first time
6. Define processes, risk scenarios & controls for the activity, based on analysis of practice; decide which controls need rules, developed & define rule users
8. Test and approve rules, incl. by eventual users & store in organization memory
7. Develop & write appropriate rules

THINKING OF YOUR OWN SAFETY AND SNITCHING ON YOUR PAL

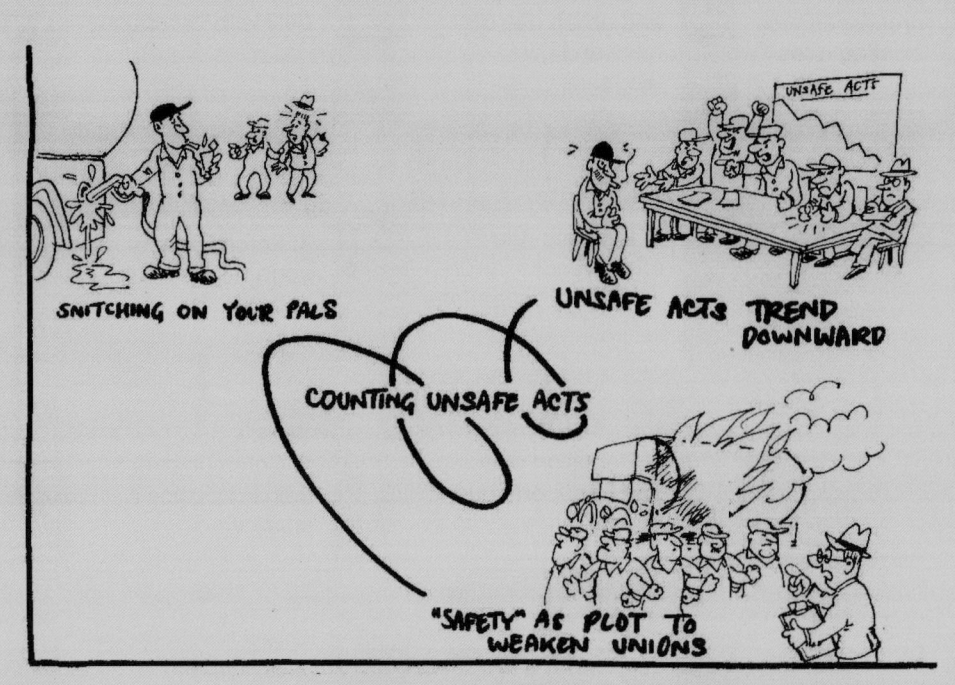

SNITCHING ON YOUR PALS

UNSAFE ACTS TREND DOWNWARD

COUNTING UNSAFE ACTS

"SAFETY" AS PLOT TO WEAKEN UNIONS

DEFENDING THE COMMUNITY OF UNIONISTS FROM MANAGEMENT SANCTIONS

SAFELY TAKING RISKS

CHARLES HAMPDEN-TURNER

At Shell, where I once consulted, they believed that you rarely learned from accidents. They were too horrible, too distressing to recall. After accidents, productivity would slow to snail's pace; people were haunted, and then would forget and endanger themselves again. You could only learn from something less distressing, an "unsafe act". If you counted the unsafe acts you witnessed and eliminated these, safety would improve.

For example, fuel tankers whose plastic shells are swept by the low-hanging branches of trees can generate static electricity. When you fill these with fuel and it splashes, there can be an explosion. It's 100-to-1 odds against it exploding and perhaps 100-to-1 against any branches having generated electricity, but with thousands of trucks filling up, it's going to happen sooner or later, and the whole terminal could be incinerated. Unless you classify it as "unsafe" and arrange for the fuel not to discharge unless the nozzle is far inside the tanker, and unless *anyone who witnesses an unsafe act immediately reports it*, nothing changes.

But would they? How do you look after your personal safety and still support your trade's union and fellow workers? Who wants to be a snitch? The term 'safety manager' is a misnomer because *everyone needs to manage safely*. Yet, if you report a fellow worker to the safety office, he will be sanctioned and management will become more punitive than ever. You aren't going to cost your mate his job or see his pay docked! The picture below shows you *thinking of your own safety and snitching on your pal* on the vertical dimension. But you could protect fellow union members by *defending the community of unionists from sanction by managers,* see the lateral dimensions and the picture bottom right. If you do that, a further accident to the one pictured is likely to occur. So, will you be hated by fellow workers (top left), or complicit in a serious accident (bottom right)?

It's hardly a happy choice.

But there was a solution possible, and we implemented it. Anyone witnessing an unsafe act should report it at once and would thereby *protect the culprit from any sanction by management.* Indeed, reporting it would *save him*, even if the safety office had already learned of the breach. However, that person must appear before a team of fellow union members, plus the safety officer, and listen to what they thought of his endangering them all (top right picture). As unsafe acts trended downwards, serious accidents disappeared, and bad habits were caught. Being unpopular with your mates is unpleasant, but it's better than being burned alive. The union asked if it could recommend dismissal for repeated offences, and this was granted. Safety is a serious issue for everyone and deserves unanimous support.

Safety has everything to do with paradox, just as we risk, yet take, precautions, so we support our community and save our individual selves from danger. The trick is to resolve this dilemma.

Professor Charles Hampden-Turner has a doctorate from Harvard and taught at Cambridge for 18 years. He is a winner of the Douglas McGregor Memorial Award and has written 22 books, including *Riding the Waves of Culture*. He's co-founder of *Trompenaars Hampden-Turner* and won Guggenheim and Rockerfeller fellowships in the USA.

CLEAR MESSAGES, COMMUNICATED EFFECTIVELY

IAN HART

I should caveat this segment before I begin. I am journalist, not a safety professional, and I have been working in the trade media for coming up on ten years now.

Because my background is not in safety, it's paramount that any message I portray to my audience is clear, concise and easy to understand. If I try to overcomplicate the message, in order to try and speak on a level with experienced safety professionals, the message can easily get lost.

It struck me that the message of safety is often quite simple, too, and the more complicated a set of rules and regulations are to follow, the less likely they are to be followed.

As a newcomer to safety, I've always been intrigued about the reputation safety has amongst the public. Particularly in the UK, which is widely seen as having a good safety record. Why is safety perceived in the way it is? Often, it's down to a vastly outdated reputation and thinking that safety is somehow there to obstruct people from going about their business. To paraphrase what former NEBOSH Chief Executive Teresa Budworth once told me, *safety is not about eliminating all hazards; it's about ensuring that you can do things which are potentially quite hazardous, safely.*

A key element in archiving that is *communication*, a two-way discussion between the person carrying out the task and the safety manager, risk assessor or business leader. Ensuring that you – whether as a safety practitioner or senior leader – are approachable, available and open to discussion eliminates one huge barrier.

Once an agreement has been made, you then need to make sure that message is effectively relayed to the relevant parties, in as simple a manner as possible in order to keep them engaged, but at the same time accurately highlighting the risks involved in the task.

Everyone has heard the safety briefing on an aircraft multiple times, but the announcement is still there loud and clear, underlining the importance of the message and how it may differ slightly from aircraft to aircraft.

That said, is everyone on that plane giving the announcement 100% of their attention? No, they're not, *because* they've heard them so many times. Creating common themes and strands to your initiatives and finding ways to keep them fresh are crucial if you want to hold the attention of those you are communicating with. Then, a clear and effective message may not just be heard, but heeded.

Ian Hart is a journalist working in Health and Safety. Hart joined Informa in 2018 as Editor, Safety & Health Practitioner. Prior to moving to Informa, he was a professional gamer and worked in business-to-business trade print media in the automotive sector.

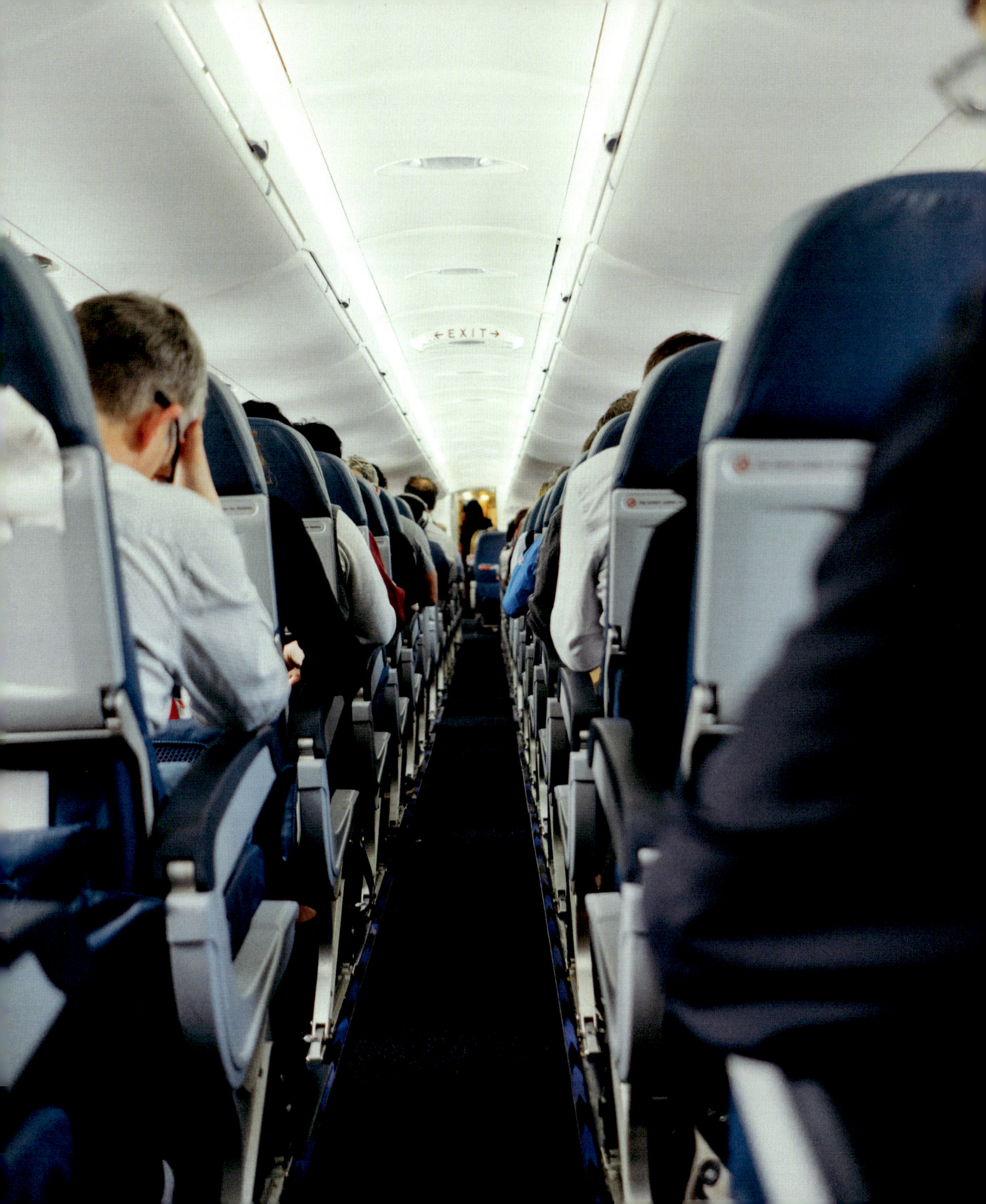
EXIT

WORLD

THE WORLD IS TEMPORARILY CLOSED

APPLYING NEW LESSONS TO THE LONG TERM

JIMMY HASLAM

The COVID-19 pandemic shook the Pilot Company, just as it did in your businesses around the world. The silver lining is that we learned to look at our business in a new and different way. In the short-term, there were a myriad of issues with which to address and resolve. It caused us to take a step back to look at how we conduct our business – particularly in the areas of health and safety – and ways we could make positive changes long-term for our team members and guests. I believe that the changes we are making in response to COVID-19 will prove to be positive innovations long-term that could be adapted to businesses in many sectors.

We believe the retail and food world will, and should, begin to use more touchless technology going forward. This will put a higher emphasis on how we conduct business using this new technology, including our Pilot Flying J mobile app, which professional drivers currently use to activate pumps, reserve showers and redeem digital coupons without touching anything besides their personal phones. We are in the process of developing similar options for our gasoline customers, in order to help our team members and guests be less vulnerable to infection and disease.

Cleanliness has always been important to Pilot Company since it was founded in 1958, but it is now more important than ever for us to provide a clean, safe and friendly environment for everyone at our travel centres. In addition to a heightened focus on providing a clean store and workplace, each guest will be personally greeted and welcomed to our Pilot or Flying J locations – physical distance does not have to equal an impersonal experience.

The vast majority of our team members work in stores located across the United States and Canada. Going forward, a much higher percentage of training will be conducted virtually. This will allow us to move more quickly and efficiently in terms of training team members. In regard to our employees who work at our support center (our corporate office), we are looking at opportunities where 10% to 20% of our team could permanently work from home. In appropriate positions, this flexibility has been shown to aid wellbeing beyond the pandemic.

As difficult as the pandemic has been in the short-term, we hope it will pay long-term benefits for the team members and guests of Pilot Company, our country and our world. By embracing technology, applying high standards of hygiene, and rethinking the way we work across industries, I believe, together, we can make the world a safer and healthier place to live and work.

James A. "Jimmy" Haslam III serves as CEO of Pilot Company. Pilot is the 10th largest privately held company in North America and employs more than 28,000 team members. Joining in 1976, he has led the strategic growth of the company, which annually supplies 11 billion gallons of fuel. Haslam is also co-owner of professional Amercian football ('NFL') team the Cleveland Browns.

AVOIDING THE UNSAFE IS NOT 'A CULTURE OF SAFETY'

LEANDRO HERRERO

Organizations have traditionally used a three-legged approach when creating a 'culture': communications, training and compliance. None in isolation have the power to shape a culture. The three together are a good 'machine gun approach' and, as such, result in lots of wasted ammunition.

Communicating about what a culture should look like, in terms of values or behaviours, is a necessary task, but it's hardly sufficient. Channels get saturated and eventually, people switch off. For example, let's take safety in an oil and gas enterprise. Communicating that safety is a goal is key. Reassuring that the entire leadership of the company is behind this drive is very important, expected and welcome. But it's hardly an engine to create a culture of safety. Look at any single oil and gas company with a safety disaster and show me that the importance of safety had not been communicated. You won't be able to.

Then, we have compliance. Rules and regulations naturally follow, particularly in territories such as oil and gas and transportation. Compliance systems explain what needs to happen and what is not acceptable. But a compliance system is mainly a threatening one. When you learn to drive, you remember the penalties for speeding more than why those speed limits and regulations are there in the first place. Compliance systems alone have little power to create culture.

Now, training. Training systems provide you with information, knowledge and skills – plus a reference to the penalties of noncompliance. Yet training alone does not create culture. It creates a well-trained workforce. The organization may become proficient, not necessarily at building a safety culture, other than… a culture of training.

The 'perfect culture' does not need much communication, compliance or training. A culture of safety is one where safety is a normal day-to-day conversation equivalent to football or soccer, or whatever conversation is around a water-cooler. If safety is an add-on, something one has to think about and 'bring in' artificially, then this is not 'a culture of safety'.

Cultures are not created in classrooms. Cultures require behavioural spread, and behaviours don't like classrooms. They like the playground. Shaping a culture requires a peer-to-peer, bottom-up system. If this exists, then training and communications top up and multiply. But the other way around is a waste. A push system (communication, compliance, training) without a pull system (behavioural, bottom-up, grassroots), is a weak system. You need strong push-pull combinations.

A reason why it's so difficult to shape culture in the areas of health and safety (oil and gas and transportation), or ethics and good conduct (banking and finance) is because 'culture programs' in those areas are entirely focused on avoiding the negative instead of reinforcing the positive: no accidents, no risk. But, for every energetically fought negative behaviour, there are thousands of positive ones never reinforced. It seems to follow the rule that complaints are actionable; compliments are not. But, for every unsafe event, there are thousands of safe ones. Reward safety, not avoidance of the unsafe.

Dr Leandro Herrero pioneered Viral Change™ an innovative approach to behavioural and cultural change in organizations. He's co-founder of The Chalfont Project Ltd, a firm of organization architects and the author of several books on leadership and management of change including *Viral Change™*, *Homo Imitans*, and *The Flipping Point: Deprogramming Management*.

CULTURES ARE NOT CREATED IN CLASSROOMS

USING INNOVATION AND TECHNOLOGY

VINCENT HO

When you already have an effective Safety Management System in place and a competent team working interdependently with a positive safety culture, your best bet to see further improvement in safety performance would be through innovation and technology. Here are some examples where we can use technology initiatives for a safer workplace. I am confident they will keep you and your workplace at least *one percent safer*.

Effective use of safety apps
It is extremely unsafe to use mobile devices while doing safety critical jobs, or even while walking, due to possible distraction. However, mobile devices are effective in reporting near-misses and hazards, sharing information in text and videos, accessing a library of work instructions, training, remote monitoring and more. If this can be combined with Building Information Modelling and diagnostic software, we can get useful information and even big data analyses at our fingertips. We need to invest more resources in safety-related apps that can significantly improve the way we work, the way we communicate and the way we process data.

Wearable devices
The latest breakthroughs in wearable electronics allow more effective devices to improve safety. Data from wearable devices such as a smart-safety helmet allows a close monitoring of workers' vital signs, whereabouts, stress and strain levels, usage and rest frequency and more. This also allows an early extraction of workers in trouble, such as early indication of heart exhaustion or fatigue, and pinpointing locations for rescue. The data collected can aid audit reports, incident investigations, competency control.

Smart sensors, Internet of Things and use of drones
Machines, buildings and engineering systems can be monitored by smart sensors and drones to provide accurate data on the 'health status' of the workplace and whether there are any accident blackspots that need to be addressed. This can reduce operations and maintenance costs while improving safety management. Drones are particularly useful in emergency response to provide an aerial view of the situation.

Design for safety
Following the principle of risk control, the most effective approach to avoid a hazard in the workplace is by eliminating or controlling it from a safe design. For example, we can eliminate 'fall from height' hazard by involving architects and engineers at the design stage of a structure to eliminate the need to work at height. If that cannot be done, a safe access should be introduced at the design stage rather than relying on a safe working routine later.

Wellbeing of employees
Employers should pay more attention to their workers' physical and mental health and wellness. Automated process and artificial intelligence can be used in areas where human errors should be reduced. Employers will take a closer look at employee lifestyle habits: their fitness, diet, smoking habits, blood pressure, cholesterol readings and the like. It reduces not only healthcare costs but also accident risks.

Ir Prof Vincent Ho, PhD, MBA, CEng, CFIOSH, FIMechE, is a risk management and safety professional with 35 years of experience in transportation, nuclear and defence industries. He promotes the applications of risk management in enhancing safety and served as 52nd President of the Institution of Occupational Safety and Health.

HOW NOW ARE YOU?

INFLUENCE THE INFLUENCERS

DAVID HOFMANN

With the rise of social media, online forums and user evaluations of just about every product imaginable, we must reconsider the roles of experts and leaders versus peers. Research[1] suggests that although experts can be influential early – when little information is available – peers and others become more influential later. Moreover, a number of research studies and books[2] suggest peer opinion is a significant influence on individual behavior.

In my research[3] assessing safety climate, one of my favorite sub-scales asks respondents the degree to which "the best" workers expect others to behave safely. This highlights the role of peer influence in modeling behavior, which I believe to be an essential aspect of a strong safety culture. For safety to truly become embedded in the culture of an organization, the key peer influencers in the organization must expect, model and hold others accountable for safe performance. As I like to ask managers and senior executives, "When you leave the room, what do the people who *others look up to* say about the importance of safety?"

How do you influence the influencers so that safety becomes truly embedded in the culture?

Identify the influencers
Individuals are not always very good at assessing their own influence, so use social network analysis and nomination processes to identify the key influencers in the network. For example, ask workers: "Who really influences the way things get done around here?" Across a sample of workers, it will become clear which names repeatedly show up.

Engage them in the development of your safety model
Engagement and involvement in the development of change efforts breeds commitment – it also ensures the procedures, practices, and protocols developed are consistent with the broader system of work design.

Make safety about protecting others (not just themselves)
Connecting one's work to the beneficiaries of that work can significantly impact prosocial behaviors whilst highlighting the ability of safe behaviors to protect others (vs. oneself) can significantly increase compliance.

Hold them accountable for their influence
Finally, organizations must create a system of accountability for these key influencers and how they are exercising their influence over others. Leaders need to ensure these individuals understand that with influence comes responsibility. It is their duty to ensure they exercise their influence to truly embed safety into the culture of the organization.

1. de Vries, R.A.J., Zaga, C., Bayer, F., Drossaert, C.H.C., Truong, K.P., & Evers, V. (2017). Experts get me started, peers keep me going: Comparing crowd- versus expert-designed motivational text messages for exercise behavior change. PervasiveHealth Proceedings. https://dl.acm.org/doi/pdf/10.1145/3154862.3154875. Flanagin, A.J., & Metzger, M.J. (2013). Trusting expert- versus user-generated ratings online: The role of information volume, valence, and consumer characteristics. Computers in Human Behavior, 29, 1626-1634.
2. Ivyngar, R., Van den Bulte, C., Eichert, J., & West, B. (2014). How social network and opinion leaders affect the adoption of new products. Marketing Intelligence Review, 3. Keller, E., & Berry, J. (2003). The influentials: One American in ten tells the other nine how to vote, where to eat, and what to buy. Free Press. Weimann, G. (1994). The influentials: People who influence people. SUNY Press.
3. Hofmann, D. A., Morgeson, F. P., Gerras, S. (2003). Climate as a Moderator of the Relationship Between LMX and Content Specific Citizenship: Safety Climate as an Exemplar. *Journal of Applied Psychology, 88*, 170-178.

Dr David A. Hofmann is the Hugh L. McColl Distinguished Professor and Senior Associate Dean for Academic Affairs at The University of North Carolina's Kenan-Flagler Business School. His research and consulting activities focus on organizational climate/safety climate, leadership, organizational change, organizational design and decision making.

#

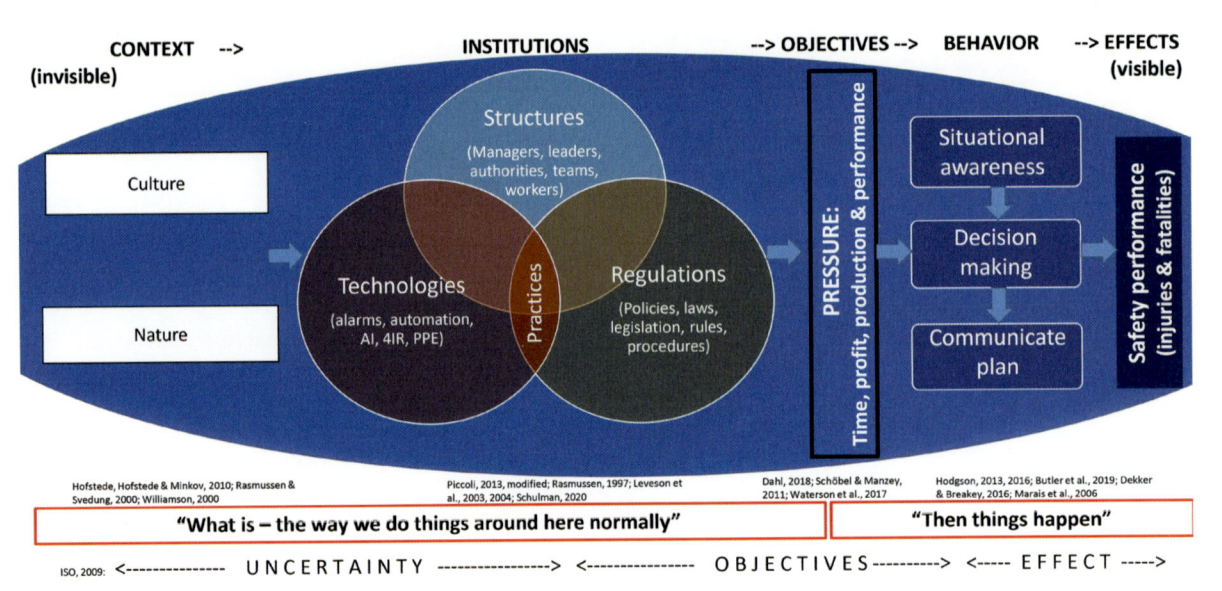

From sociotechnical system causes to safety outcomes

CONTEXT -->   **INSTITUTIONS**   --> **OBJECTIVES** -->   **BEHAVIOR**   --> **EFFECTS**
(invisible)   (visible)

- Culture
- Nature

Structures (Managers, leaders, authorities, teams, workers)

Technologies (alarms, automation, AI, 4IR, PPE)

Practices

Regulations (Policies, laws, legislation, rules, procedures)

PRESSURE: Time, profit, production & performance

- Situational awareness
- Decision making
- Communicate plan

Safety performance (injuries & fatalities)

Hofstede, Hofstede & Minkov, 2010; Rasmussen & Svedung, 2000; Williamson, 2000   Piccoli, 2013, modified; Rasmussen, 1997; Leveson et al., 2003, 2004; Schulman, 2020   Dahl, 2018; Schöbel & Manzey, 2011; Waterson et al., 2017   Hodgson, 2013, 2016; Butler et al., 2019; Dekker & Breakey, 2016; Marais et al., 2006

"What is – the way we do things around here normally"   "Then things happen"

ISO, 2009:   <--------------- U N C E R T A I N T Y ---------------->   <---------------- O B J E C T I V E S ---------->   <----- E F F E C T ----->

SAFETY, CULTURE & SOCIO-TECHNICAL SYSTEMS

GERT JAN HOFSTEDE & FRED GOEDE

Safety management is regularly confronted by hazards from nature, such as tsunamis, adverse weather-related events and pandemics, in addition to workplace hazards. In response, organizations and safety management in particular, have relied for decades on formal structures. These include increasing regulations, technologies and leadership structures. Despite these efforts, some of the largest environmental and safety incidents occurred in the last decade, and the goal of zero fatalities remains out of reach. An investigation into major safety incidents shows that 70-80% of reports blame the 'wrong safety culture', 'wrong behaviours' or 'non-compliance'. This underlying culture seems important to successfully implement our conventional safety management interventions.

The coronavirus outbreak provides an excellent case study to demonstrate how different cultures around the world respond differently to the same threat. On a smaller scale, organizations are challenged to implement safe working conditions and abide by regulations and procedures, adopt new technologies and ensure adequate supervision. It is unlikely that these interventions will be successful without a better understanding of the safety culture in the workforce.

Our conceptual model is based on 25 years of industry experience, to which is added a systems perspective. It describes what happens when nature throws its hazards at us and how underlying culture determines the organizational response. We highlight the opportunities for safety improvement in the dynamic workplace where workers and their organizations are pressured to perform. We believe that better understanding of how workers and their organizations self-organize within their cultural setting assists us to improve safety performance.

Start on the left of the figure, where the ordinary situation is depicted. At far left we see *context factors*: culture, as the propensity of people to behave in certain ways, including unwritten rules of the relationships within the workplace; and nature, in the form of *hazards*. These impinge on the entire system (the large blue oval). We see the *socio-technical system* proper (centre): practices in the organization, determined by three things: *technologies*, *formal structures*, and *regulations*.

Then things happen: there is an explosion, an epidemic... We first see (at right) *pressures*, e.g. for performance, that the socio-technical system exerts over time, which could push its resilience to its limits; and resulting behaviours: *situational awareness, decision making* and *communication*. At far right there's the effect in terms of *safety performance*.

This model allows discussion of elements that contribute to safety outcomes, both after the event and in order to prevent new occurrences. It can serve as the basis for constructing agent-based simulation models that facilitate understanding of the role of the various system elements over time. Such models are miniature worlds in which coincidences can happens much as they do in industry.

Gert Jan Hofstede is professor of Artificial Sociality at Wageningen University, the Netherlands. Educated as a population biologist, he was a computer programmer before entering academia. Hofstede co-authored the best-selling *Cultures & Organizations: Software of the Mind.* He's involved in modelling human social behaviour in the context of socio-technical and socio-ecological systems.

MORE OR LESS?
ERIK HOLLNAGEL

More or less?

"If we can make the world just 1% safer, then 28,000 lives will be saved each year."

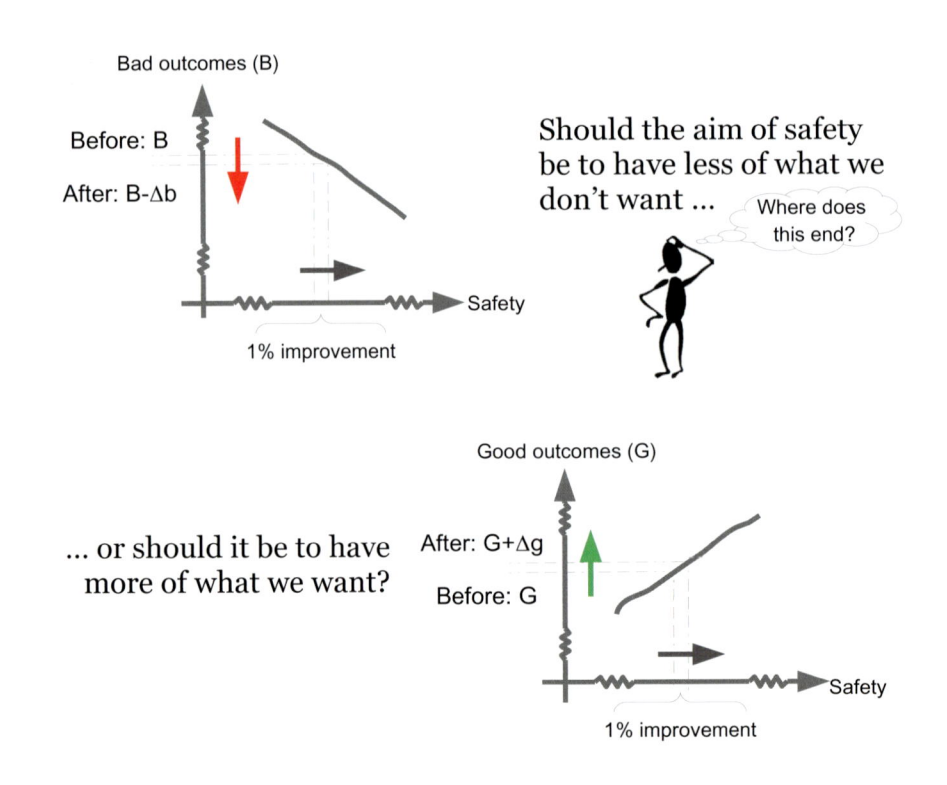

Bad outcomes (B)

Before: B
After: B-Δb

1% improvement

Safety

Should the aim of safety be to have less of what we don't want ...

Where does this end?

... or should it be to have more of what we want?

Good outcomes (G)

After: G+Δg
Before: G

1% improvement

Safety

Accentuate the positive!

Dr Erik Hollnagel is Senior Professor of Patient Safety at Jönköping University, Sweden. He has worked at universities, research centres, with industries in many countries and with problems from many domains. Hollnagel has published widely and has authored / edited 25 books, in addition to papers and book chapters.

RISKY REWARDS
ANDREW HOPKINS

"The CEOs of many large companies are paid huge bonuses based purely on the share market performance of their company. We need to ensure that safety, particularly with respect to major accident risk, is incentivised in a similarly powerful way."

Dr Andrew Hopkins is Emeritus Professor of Sociology, Australian National University, Canberra. He has authored many books on safety, including case studies of disasters in the oil and gas and coal mining industries. His writing focuses on the organizational causes of accidents, with more than 100,000 copies sold.

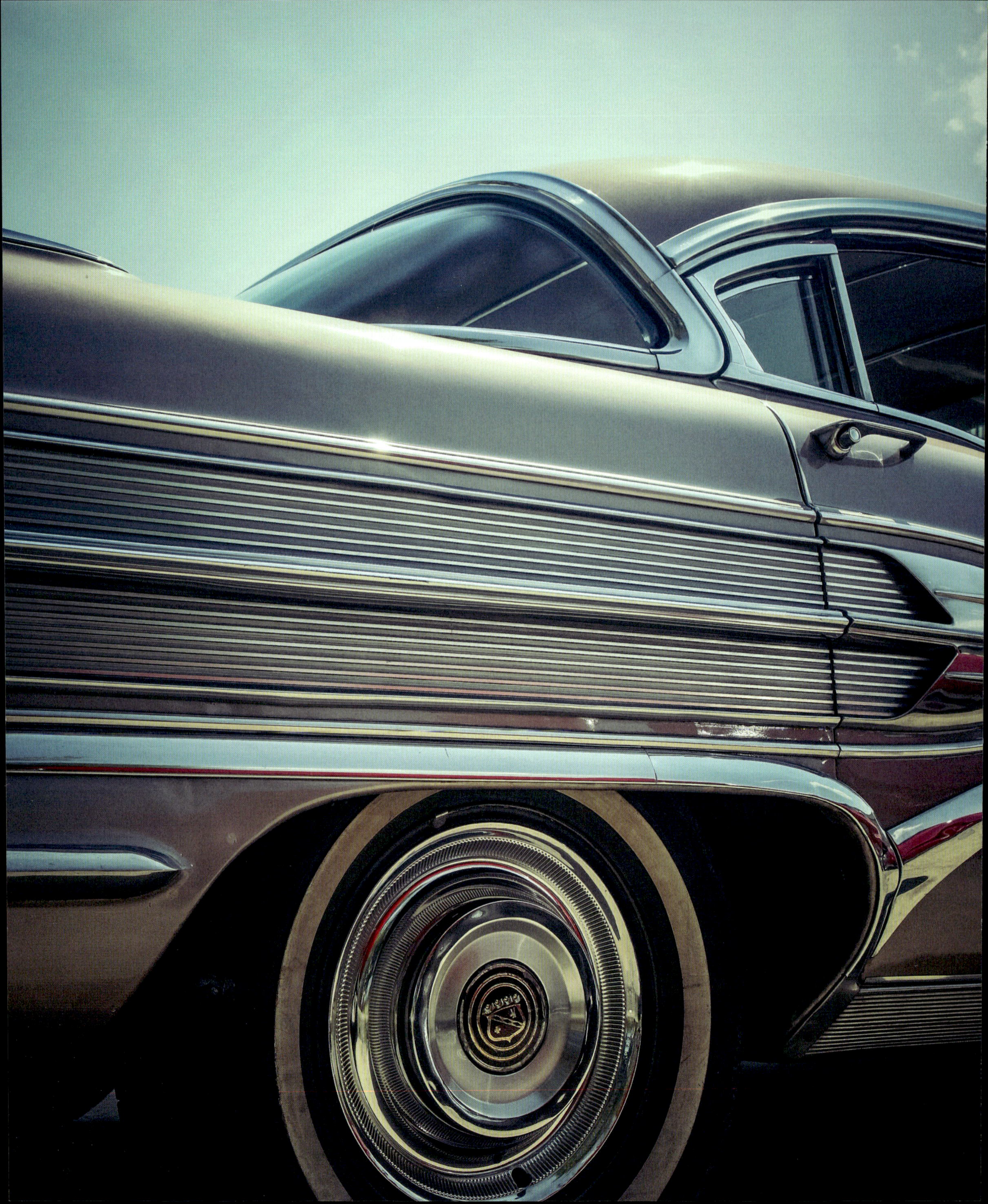

THERE IS A SILVER BULLET

PATRICK HUDSON

I've studied hundreds of major accidents and read many official reports and judicial reviews and have come to the conclusion that there really *is* a magic remedy.

If the challenge is to reduce fatal accidents and serious incidents by one percent, then this may even do a little better. The greatest single failure of high hazard organizations is that management, especially senior management, already knew what should be done, but failed to ensure it was carried out and failed to assure themselves that action had been taken. They knew what to do, from previous incidents or from the results of risk management and planning processes, but failed to implement. What they *should* have done varies a lot, but examples include journey management, making fall arresters easily available, or some process that could prevent an accident. All too often we read: "There was always something more important to be done first".

How did things go wrong? What do we need to fix? The majority (over 50%) of the several hundred fatal accidents were caused by non-compliance with established processes and procedures. This is not surprising; a common approach to improving safety performance involves making the workforce more compliant. This would seem to be the silver bullet, but for the fact that we went further and asked *why* people, typically the victims, were not complying. The answer was even clearer; people predominantly broke the rules or failed to follow procedures because their boss, and up the line their managers, were themselves not in compliance with the company's policies and procedures, and the managers were almost exclusively (over 90%) doing this for the apparent benefit of the company, not themselves personally.

Managers, especially senior executives, are always in search of a simple solution. The problem is that the power to implement the solution lies in the hands not of the immediate victims, but of those who demand the solution. Why, then, don't managers just do the right thing? Firstly, they often fall into the trap of blaming the victim, not going deeper into the reasons why they did what they, fatally, did. But there are some other reasons why this might happen. One is that, having successfully ascertained the underlying causes, it's too easy to feel the job is done and move on. There is little apparent benefit or desire to fix 'old' problems. Promotion in many organizations all too often rests on making new plans for others to implement; there is little glory or status in fixing old problems.

Many technical organizations value planning above implementation, strategy above tactics, but it's the tactical implementation that kills people. Values, attitudes and motives are laudable, but the most effective intervention is for top management to ensure people do what has been agreed and assure themselves that it has happened.

Is there a silver bullet? Yes, here it is. The only problem is that it is *they*, and not their workers, who have to make the bullet and, when the time comes, shoot it.

Patrick Hudson is professor emeritus of the Human Factor in Safety at Delft University of Technology. He was project leader of Shell's program leading to the 'Swiss Cheese Model'. As a member of the team that developed Shell's approach to safety management systems after Piper Alpha, he led Shell's *Hearts & Minds* program.

EVOLUTION BY DEVOLUTION

JON HUGHES

Evolving your OSH management system by devolving responsibility is a strategy that will make a significant positive difference, both to the effectiveness of your health and safety management systems and your organization overall.

The days of the OSH professional being the 'go to' person for all things OSH related are, thankfully, gone. The poor, maligned H&S manager was the person responsible for everything safety-related (we hadn't really looked at health back then). They were the ones writing the policy and risk assessments; selecting and ordering PPE; arranging for statutory tests; training people; investigating accidents; dealing with enforcing authorities, and more. As long as 'the boss' was kept out of court, all was well.

The problem was that the people creating the risk, those working at the 'sharp end', were mostly excluded from the process. Equally, the leadership teams were almost exclusively absent from the world of OSH, beyond signing a policy once a year. Therefore, there was a massive disconnect between what *should* have been happening – what the policy actually said – compared to what was *really* taking place. Responsibility was not sitting in the right place.

If anything here looks familiar in your organization today, don't worry. There is a clear way out of that situation in to a new way of working.

To evolve your OSH management system, the organization needs to devolve responsibility and accountability to those closest to the operation. The leadership team also needs to take accountability and shape an organization-wide vision. They will likely need some coaching and guidance on this. Once they 'get' health and safety, they need to be facilitators and 'encourage, engage and empower' (Remember those 3 E's!) their managers to take responsibility for OSH.

So what does that look like?

In one organization I know, there are 940 managers in the UK alone, and *each one* is responsible for OSH, along with many other things. They are still responsible for the projects they lead, but they're also accountable for the health and safety of themselves and anyone affected by the activities of their team. They complete the risk assessments; brief their team; provide information and instruction; report near misses and incidents. They act as conduits for information flows from the team up to the very top.

By making managers accountable and responsible, and providing them with effective training and the tools they need, as well as encouragement, engagement and empowerment, these managers will become your OSH managers too. Rather than having just one OSH professional in your organization, you will know have a much larger team of people, working in concert. Your OSH professionals can then devote time to developing strategies, providing assurance to the leadership team, mentoring colleagues, and be less involved in the minutia of managing OSH at the sharp end.

If you can crack that, then you'll have an excellent management system on your hands, one that has evolved by devolving responsibility and accountability to those best placed to manage it.

Jonathan Hughes CFIOSH is the Global Lead for Health & Safety and Business Management Systems at international consultancy business, Turner & Townsend. With 18 years' experience in managing and leading health and safety, in house, and working with clients, organizations value the pragmatic and practical advice that Jonathan provides.

CREATING SAFETY THROUGH ENGAGED LEADERSHIP

STUART HUGHES

So, you lead an organization or a team, congratulations … but do you lead safety? Creating safety *is* leadership and committed leadership *produces* safety.

What do I mean?

Have you ever said: "Safety is our number one priority"?

Stop it, and stop it *now!*

Catchy slogans are meaningless in an organization as soon as something conflicts with them. This happens when plant modifications designed to improve safety aren't approved, or employee feedback on how to make safety improvements go unactioned. You – and your slogan – lose all credibility.

How then, as a leader, do you create safety within your organization? Intent, alignment, commitment, language and action. These five areas, when implemented well, create safety and deliver performance.

Intent: I often hear leaders say, "Safety professionals don't understand business and their organizational objectives". Sound familiar? Well, it shouldn't. Not if your organization has a clear intent. Ask yourself: Do you know the intent of your safety team/advisor? If the answer is "no," there is your first problem. If your safety team/advisor doesn't understand your organizational intent, and you don't understand theirs, then you're out of alignment.

Alignment: Does safety feature within your mission, vision and values? If it doesn't, then are you really displaying a commitment to safety? Aligning your language through your mission, vision and values to include safety is an easy step to demonstrate a commitment to creating safety.

Commitment: Are you leading by compliance? Do your employees choose to work safely or are they simply working in compliance? When we operate solely in compliance, we remove the potential for developing improvements in safety. Groups do not develop commitment. Individuals do. Therefore, you need safety to become a personal commitment. As a leader, do you have a personal commitment to safety? If not, develop one. A simple sentence committing to a safety action is a great place to start.

Language: Does your language close down safety discussions or open them up? Instead of "I'm going to launch this new process next week focused on safety, let me know what you think." Try: "I've created a new process designed to improve safety. I know there are people in this group with additional experience and knowledge to mine. Therefore, you could identify ways to improve it. I'm looking to you to challenge the document and suggest improvements." See which gets more engagement and feedback from your team!

Action: It is your actions as a leader that have the greatest impact. Look in the mirror rather than through the magnifying glass. Identify the action you can take, demonstrating your personal commitment to safety. Have you ever publicly praised a line manager for completing their safety inspections on time? If not, try it and watch the change in your next review of safety inspections. Once this is in a good position, choose another area. Say, actioning employee feedback. Over time you will see the impact of this and witness others personal commitment to safety in action.

Stuart Hughes is the Head of Health & Safety for Mercedes AMG PETRONAS FORMULAONE TEAM, current World Champions, having won six consecutive double world championships. Hughes is a safety thoroughbred with a 15-year career dedicated to Occupational Safety & Health (OSH), specializing in enhancing OSH performance and driving cultural change.

HOW ARE YOU, REALLY?

MENTAL HEALTH: NOW IS THE TIME TO ACT

ROD HUNT

How do we save the lives of an additional 28,000 people in the workplace each year, globally? We challenge businesses and their leaders to renew their commitment to mental health in the workplace and invest in solutions for a happy and healthy workforce. Peter Cheese, Chief Executive of the Chartered Institute of Personnel and Development, believes that properly addressing mental health "starts with awareness at the top of organizations".

Will addressing mental health actually save lives? The short answer is "Yes". The sad reality is that the ultimate human cost of poor mental health is loss of life through suicide. One of the starkest examples of this was seen in France recently, following the suicides and attempted suicides of 31 employees linked to the restructuring policy of a company – which also resulted in the prosecution and conviction of three senior executives.

As I write, it's Mental Health Awareness Week in the UK, and health experts at the United Nations have warned of a COVID-19 related mental health crisis and encouraged governments to put the issue "front and centre" of their responses.

Whilst historically, there has been a focus on safety and physical health, regulators have started to turn their attention to the 'unseen issue' in the workplace – mental health. Rightly so.

UK statistics for 2018/2019 show:

- 602,000 workers suffer from work-related stress, depression or anxiety
- 12.8 million working days lost due to work-related stress, depression or anxiety
- 44% of all work-related ill-health cases are work-related stress, depression or anxiety
- 54% of all working days are lost due to ill-health caused by work-related stress, depression or anxiety

The Stevenson/Farmer report found there is "a large annual cost to employers of between £33 billion and £42 billion," with over half of the cost coming from presenteeism (when workers are less productive due to poor mental health), and studies suggest this is increasing.

Since a business' workforce is its biggest asset, properly addressing mental health is a win-win. Beyond protecting your employees' welfare, benefits also include fewer accidents, lower absenteeism, higher staff retention, productivity, brand protection and your workforce going home safe and well. Mental health in the workplace is also becoming a bigger issue for investors, and it might not be long before it even appears in tender questionnaires.

If the carrot isn't enough to entice you, it is worth reflecting that regulators have enforcement powers requiring businesses to comply with their legal duties to adequately manage work-related stress. Failure to do so could result in an improvement notice or prosecution. Whilst I'm not aware of any prosecutions in the UK to date regarding work-related stress, it will be interesting to monitor how the regulator responds to the impact of COVID-19 on mental health in the workplace.

Australia is well ahead of the UK on tackling mental health, and we are seeing an active response from French authorities. This is a global issue for business which, unless addressed, is only going to get worse. *Now* is the time to act.

Rod Hunt is a partner in Clyde & Co's SHE Regulatory Department which has "strength in depth from partner to associate" and is retained by FTSE 100 and global businesses. Highly rated as one of the UK's leading OSH lawyers, Hunt "stands out as offering commercial, sensible and accomplished advice".

BREAKING THE CHAIN OF PAIN

EHI IDEN

Health and Safety Management Systems should look more closely into the concept of 'duty of care' of employees. This is becoming a phrase that we reference in our health and safety policies, but often isn't adhered to in practice. It is important to rethink the priority placed on the health and safety of our employees and contractors within our workplaces, and this responsibility falls to leaders.

The duty of care should go beyond just the employees and be offered to employees' families, who are, by extension, a part of the business. The death or an injury of an employee adversely impacts the employee's family, thus our organization's health and safety decisions must take these families into consideration. This is what I refer to as an 'employee and employee's family-centred health and safety approach' to business.

Every death of an employee has multiplied effects on family members, and when organizations do not understand this, we unknowingly continue to build a chain of pain, which causes people to emotionally and psychologically suffer for actions they never participated in. In an employee and employee's family-centred approach to health and safety, we do not just train the employees, we motivate the employees to train members of their families as they return home from work. With this, we are also expanding the circle of impact of our behavioural change strategies with the overall aim of keeping employees and their families safe and healthy, reducing down time and increasing productivity. We must bear in mind that accidents in employees' homes or ill health of their family members could be a source of their absenteeism from work and poor level of productivity, but this can be controlled if our management system can positively influence the safety behaviour of employees' families as third party stakeholders in our businesses.

We must create a system that acknowledges the cost (direct and indirect) of accidents or deaths of an employee not just to the business, but also for families.

What was the cost of educating that employee? How many years has the employee worked? What is the average years such employee would have been in active employment? How much of improvement in quality of life would this have brought to the family? How much pain has the workplace accident or death caused for the family? What impact has had this on the mental health and wellbeing of the family members? We should be asking these and many more questions in our effort towards breaking the chain of pain for employees and their families.

We should be reviewing our workplace health and safety management system to, by extension, accommodate the safety, health and wellbeing of the employees' families? We should have a workplace system that protects the employee and the employee's family, where the employee can feel safe, protected and able to return to work to be productive. This means increasing both profit and a conscience to the business today and long term.

Ehi Iden is Chief Executive Officer at Occupational Health & Safety Managers, Nigeria, Regional Chair, Patient Safety Movement, West Africa, on the Steering Committee for European Network Education and Training in Occupational Safety and Health (ENETOSH), President of OSHAfrica and a member of the Expert Working Group on OSH for Africa Union.

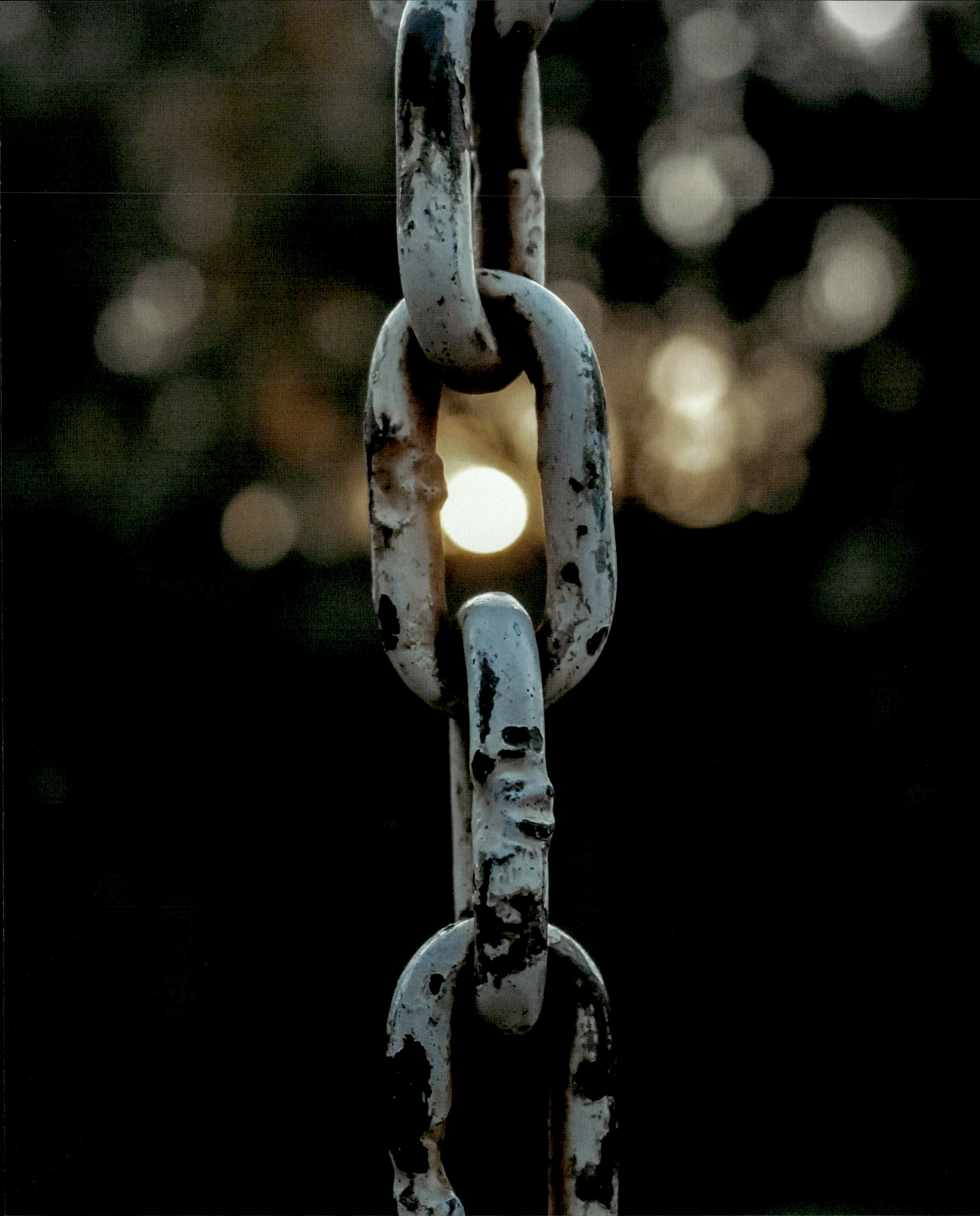

A QUESTION OF LEADERSHIP

MIEKE JACOBS AND PAUL ZONNEVELD

For a long time, a leader's ability to direct and drive results was seen as a critical trait for success. However, applying linear thinking and solutions to a complex problem can aggravate the problem, leading to unintended consequences. Our old leadership behavior presents the risk of overlooking the real issue and fixing the wrong problems.

Organizations are complex living systems. Nothing works in isolation: all elements in an organization are connected to each other. If we want to understand what is disrupting our daily operations or what complex combination of decisions led to a safety incident or accident, we need to explore the entire system and the underlying dynamics, instead of just fixing the symptoms.

What are the new leadership questions to ask ourselves, in order to safeguard longer-term safety performance and sustainability?

What will be the consequences of my decisions for generations to come?

Many leaders are driven by leaving their legacy behind. They are eager to put their stamp on the organization and to do better than before. However, leadership should not be a heroic solo effort. Incidents that happen now are often the result of decisions taken a long time ago, or a pile-up of actions. Creating a sustainable safety vision requires you to see yourself as one link in a lineage of leaders that spans generations, and to consider the longer-term impact.

How can I invite the opposing, non-confirming and muted voices?

Safety and corporate social responsibility spans across all layers of your organization and beyond its fences. Many of the industry challenges that we face require cross-fertilization of ideas, bottom-up engagement and collaboration with new partners. It requires us to admit that we sometimes simply don't know. That the solution to this question is beyond our current reach. We need to suspend our individual analysis for more collective solutions. How can we invite the wealth and wisdom of the people closest to the action, the dissenters, or even competitors and regulators?

Do I stand by my values?

The state of our planet requires many industries to do some soul-searching and take responsibility for a zero-harm vision. We can't succeed if we let ourselves be defined by the group norms and by a single focus on profit and shareholder value. We must let our own values lead the way and explore what culture we have created with our past approach. Most organizational systems have a set of unwritten rules that define how to belong. The ones who don't follow get excluded. Strong loyalty leads people to go along with the unwritten rules, even when they aren't in line with their values. By doing so they put themselves and others at risk, even with a good intentions. Now is the time to bring these to the surface and invite a culture of transparency.

These deep questions will support you in a significant step change: the one from "doing the things a leader does", to *being one – and being you.*

Mieke Jacobs and **Paul Zonneveld** are transformational facilitators, experts in systemic intelligence and co-authors of their, *Emergent*.

LEADERSHIP: RELATIONAL NOT RATIONAL

PHIL JAMES

When the COVID-19 pandemic hit in 2020, the world suddenly rediscovered just how connected people are. Viruses spread. Any of us might have the virus in our bodies, but it's the contagion that concerned us most – an invisible transmission with unpredictable consequences. A bitter reminder that we're not isolated, atom-like from each other, but connected, sometimes in puzzling, invisible ways.

What also spread were new ways of behaving in response to the pandemic that stuck with surprising rapidity. Some of these were called for by authority figures, but they took hold only because of the small changes we all made to what we were doing, oxygenated by mass social media. Ensemble applause for key workers; crossing the street to avoid a fellow pedestrian; patiently queuing in the supermarket. All of these new expectations were springing up as the new way of doing things around here. This was not organized by some great mind in authority with their hand on the controls of the system, but as a result of the little things we all started doing, and repeated, as a connected society of inter-dependent people.

Something else happened, too. Supermarket floors became striped with 2-metre chevrons of tape, and the rules were laid down about how many times we were allowed to walk the dog. Some of those things helped to create the new norms. But some folks broke the rules, held parties, came too close in the vegetable aisle or popped around to see Mum and Dad. They broke the rules in spite of the risks to their health and that of their loved ones. Why? Well, people do. Despite the guidance, leadership and social pressure, people don't behave rationally or predictably. At least not all people, all the time.

Leaders are confronted with this conundrum on a daily basis. Yet, we're meant to lead, to take responsibility for the lives and limbs of our colleagues. So, what can we do if we can't control what people do? A clue, for me, lies not in *what* to do, precisely, but in *how* we think about leadership. We have to approach leadership relationally, not rationally. And that means accepting that we're not always in the lead. In the end, even the best strategies and initiatives won't convey how much you care nearly as well as an honest conversation with a fellow human being. Giving someone space to say "no"; following through on your promises; spending an extra five minutes to find out what's on someone's mind – and sharing what's on yours. All of these will power more change than the best suggestion scheme or communication campaign that sets out the objective facts about risks to health.

Stop hiding behind your job title, and treat people with relational respect, not rational instruction. Be kind. Trust. Marvel at what can happen when you're part of something you have no control over.

Phil James is CEO of the Institute of Leadership & Management, whose mission is to *Inspire Great Leadership – Everywhere*. James holds a Masters in Corporate Direction and is a doctoral candidate at the University of Hertfordshire, studying the application of complexity science as an analogy for management and organizations.

To Plan: use the matrix top-down To Reflect: use the matrix bottom-up React to the systems		In what ways do people interact with their surrounding systems?			
		React + Adapt to the systems	React + Adapt + Change the systems to fit ourselves	Theorise about ourselves and the systems	
Parameters	What do people need to know?	What, When, Where, Who, How			Why
Parameters	When is this applicable?	-Invariable conditions -Predictable outcomes	-Some variable conditions -Some unpredictable outcomes	-Many variable conditions -Many unpredictable outcomes	Always
Actions	How could you facilitate it in your organization?	-Rules and procedures -Training -Equipment -Calculated operational time	As the cell on the left PLUS -Promotion of self-knowledge -Coaching -Peer-reviews -Appreciation of diverse knowledge, experiences, backgrounds and expectations -Plan for some more time	As the cell on the left PLUS -Provision of knowledge about the system -Education -Peer-sharing -Visible inclusion of diverse knowledge, experiences, backgrounds and expectations -Plan for even more time	-Welcome of curiosity and questioning -Provision of answers and/or sources -Dedicate time
Prerequisites	How much energy investment do you need?	Baseline investment to enforce rules and procedures top-down	Moderate investment to use bottom-up knowledge	Increased investment to design together with the end-users	Variable to offer meaningful answers to questions
Prerequisites	What are your success ingredients?	Visible management commitment – Multidirectional and psychologically safe communication – Mutual trust			
Results	How might your system look like?	Arranged to deal with predetermined tasks and outcomes	Arranged + Flexible to deal with disturbances in tasks and outcomes	Arranged + Flexible + Inventive to improve tasks and outcomes	Balanced in Doing, Thinking and Feeling alike
Results	What might your system gain?	High Consistency	High Consistency + Mindful Adaptivity	High Consistency + Mindful Adaptivity + Informed Innovation	Appreciation and Engagement
Caution	Where should your energy be directed, and the gains be realised?	Equitably across all system properties and objectives (safety, security, quality, efficiency, productivity, etc.), unless temporary redirection of resources is justified and return to equity is secured.			

WHAT DO YOU WANT TO LEAD?

NEKTARIOS KARANIKAS

Sooner or later, most of us realise that one-size-fits-all approaches do not work across different (sub)systems. However, what does this really mean for you?

You can use the matrix to the left to plan actions depending on basic system parameters and desired results after securing the respective prerequisites.

Also, you can consult the matrix when your system does not behave as expected, and you wish to reflect and discover reasons. The more the matrix cells are vertically aligned, the more proper and effective your leadership and management.

Nektarios Karanikas is Associate Professor in Health, Safety and Environment at the Queensland University of Technology, Australia. Before switching to academia, Nektarios served 19 years as a military officer in the Hellenic Air Force, in various positions related to aircraft maintenance as well as safety and quality management.

DO YOU LIKE THIS SAFETY THING?

TRISH KERIN

Storytelling is a valuable way for humans to learn lessons and pass on knowledge. Fun rhyming poetry acts in a similar way.

This poem outlines what I believe are key leadership behaviours needed to improve safety.

Do you like this safety thing?
Or are you not sure what to bring?

Let's talk about your leadership,
On what do you have a grip?
It's all important to be yourself,
And consistency is just top shelf!

Of information, you must seek,
Learn from others, stickybeak!
You must decide which way to go,
Sometimes you'll host a speaking show.

Always make time for your reflecting,
And remember, people need protecting!
Your workload may be big to run,
But never forget to have some fun!

Do you like this safety thing?
I do! I have so much to bring!

Trish Kerin is the director of the Institution of Chemical Safety ('IChemE') Safety Centre, a leading international process safety body. An engineer by qualification, her strengths are leadership and organizational culture. In 2018, she received the Trevor Kletz Merit Award from the Mary Kay O'Connor Process Safety Center.

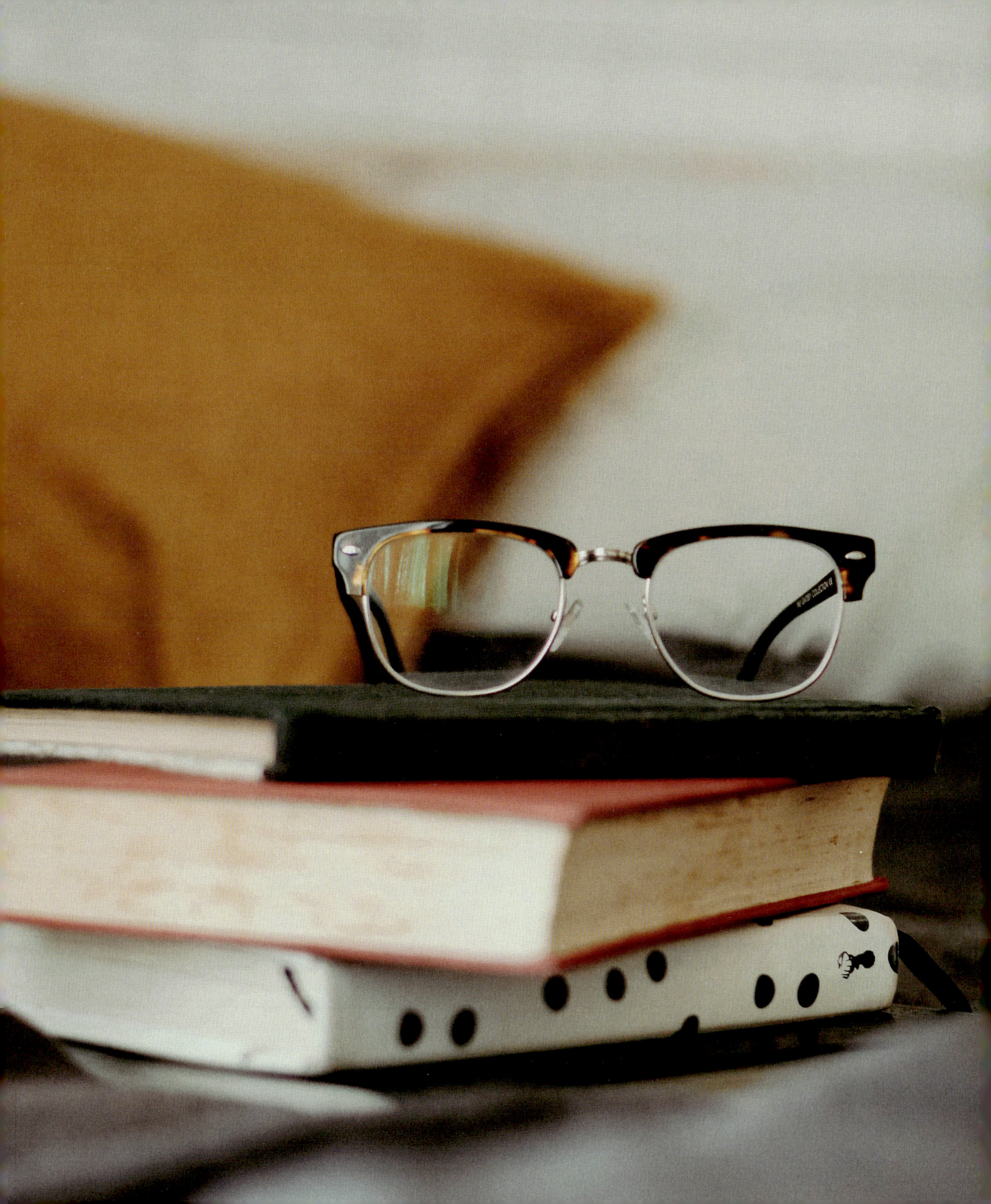

LET US KEEP UP STANDARDS

THE FUTURE OF WORK

EVELYN KORTUM

I kept wondering why we cannot get a grip on stress at work. Why was the situation not improving? We have plenty of research on how to prevent and address work-related stress from an organizational, workplace and individual level. And the problem keeps persisting – it even felt like it was getting worse instead of better. This was despite the fact that mental health issues finally had been gaining momentum through several high-level initiatives. The UN Secretary General launched his Mental Health Strategy in October 2018, and the WHO is developing guidelines to address mental health at work, among others.

Then, the public health crisis and pandemic of COVID-19 hit. For many, the kind of stress we had been studying ceased to exist due to working from home. Some people even loathe to go back to work and be exposed to the same untenable situation as existed before the crisis. For others, the situation worsened, because the stress of survival and keeping one's job suddenly became prominent. Those of us who have a stable income are privileged, while most of the world's workers do not have the stable income and the health and safety net that many of us enjoy.

The 'new normal' is being talked about, but no-one knows exactly what it means. Maybe it's more teleworking from anywhere in the world, or more precarious jobs, higher flexibility? Decline of existing health and safety nets? That working in the way we know it will change is clear to all of us. The ILO reports and other reports on the future of work are certainly giving us a taste of what might come.

Hence, the future of work has started. Even the 2020 World Health Assembly met online, which was unthinkable before this public health crisis. Other large fora, like the UN General Assembly, are considering meeting online meeting. This will mean less stress of traveling to places and less air pollution, among other things. Eventually, there might even be a long-lasting positive impact on our environment. The future of work will most likely be a hybrid of before and after COVID-19. However, changes are imminent, whether we go back to our workplaces for some or many days of the week, with or without masks, perhaps avoiding physical closeness, and so on. This crisis has taught us that we can work from home and how this might even increase quality of work and family life.

Many of us miss our colleagues. Seeing and talking with colleagues face-to-face is still relevant and important, owing to our human nature. There are always downsides to the good things in life. Creativity has certainly been heightened for many, and we've had time to think. Time we didn't use in commuting to work is time we can do other things.

2020 is definitely a year to remember for humankind. Let us keep up standards in all spheres of work, life and the environment, more safely, and perhaps, more serenely.

Dr Evelyn Kortum is an OSH expert seconded from the World Health Organization currently serving the Federation for International Civil Servants Association as General Secretary. She is a member of ICOH, APA and BPS with a PhD in Applied Psychology and a Masters in Occupational Psychology.

A WORKER WHO DOESN'T KNOW HOW TO DO THE JOB CAN'T DO IT SAFELY

PHIL LA DUKE

Probably the easiest and most impactful change we can make on occupational safety is working – and meeting – from home. While in the US, commuter injuries are not considered workplace injuries, they are in many other parts of the world. Statistically, you are far more likely to die during your commute to or from your workplace than while physically on site. The pandemic demonstrated that many jobs can be done from a person's home without a loss of productivity or degradation of quality, *and* it good for the environment. That said, I'm not sure the world is quite ready for that, so I'll offer the following instead.

In the entire histories of Safety and Industrial Learning, there has been a major gap between the two disciplines. Competence, that is a worker's ability to perform the tasks critical for operating at minimal risk, is often ignored by safety and the learning functions. When you are teaching someone how to be proficient at a (potentially risky) job, there is no room for learning by trial and error.

Far too often workers are provided dull, poorly-designed training that meets a regulatory requirement but imparts no real skills, knowledge or wisdom. What's more, studies have shown that very little of the training provided is retained and ultimately practiced back on the job. Take 'shadow training,' for example, when an experienced worker is tasked with showing the new worker how to do the job. The veteran often is eager to move on to his or her new assignment and may be impatient with the recruit. The worker receiving training, for his or her part, is eager to please and demonstrate competence. The entire exercise results in about 20% retention.

The learning function doesn't want to touch 'safety training,' for fear of liability or simply getting it wrong—despite regulations that seem to be more about attendance at an ill-defined event than about true competency building.

Competency Assurance must be hard-wired into how businesses approach worker safety. I often say, "Nobody wants to get hurt, and your process isn't supposed to hurt them" and "If a person can't do the job, he or she can't do it safely."

I want Continuous Improvement teams to bring rigorous investigative methodologies and tools to the Incident Investigation process so that 'worker error' stops being the default root cause of so many injuries. Additionally, I'd like to see this discipline introduce multiple causations to the Safety function. Often, an injury is not caused by a single root cause, but by many, into related failure modes. Finally, I'd like to see periodic performance evaluations of the workers' competence – today's performance evaluation is more about attitude and behaviours, instead of the proficiency with which people are able to do their jobs.

Phil La Duke is an OSH speaker and writer with more than 1,000 works in print including *I Know My Shoes Are Untied! Mind Your Own Business, Lone Gunman: Rewriting the Handbook On Workplace Violence Prevention* and *Blood In My Pockets Is Blood On Your Hands.*

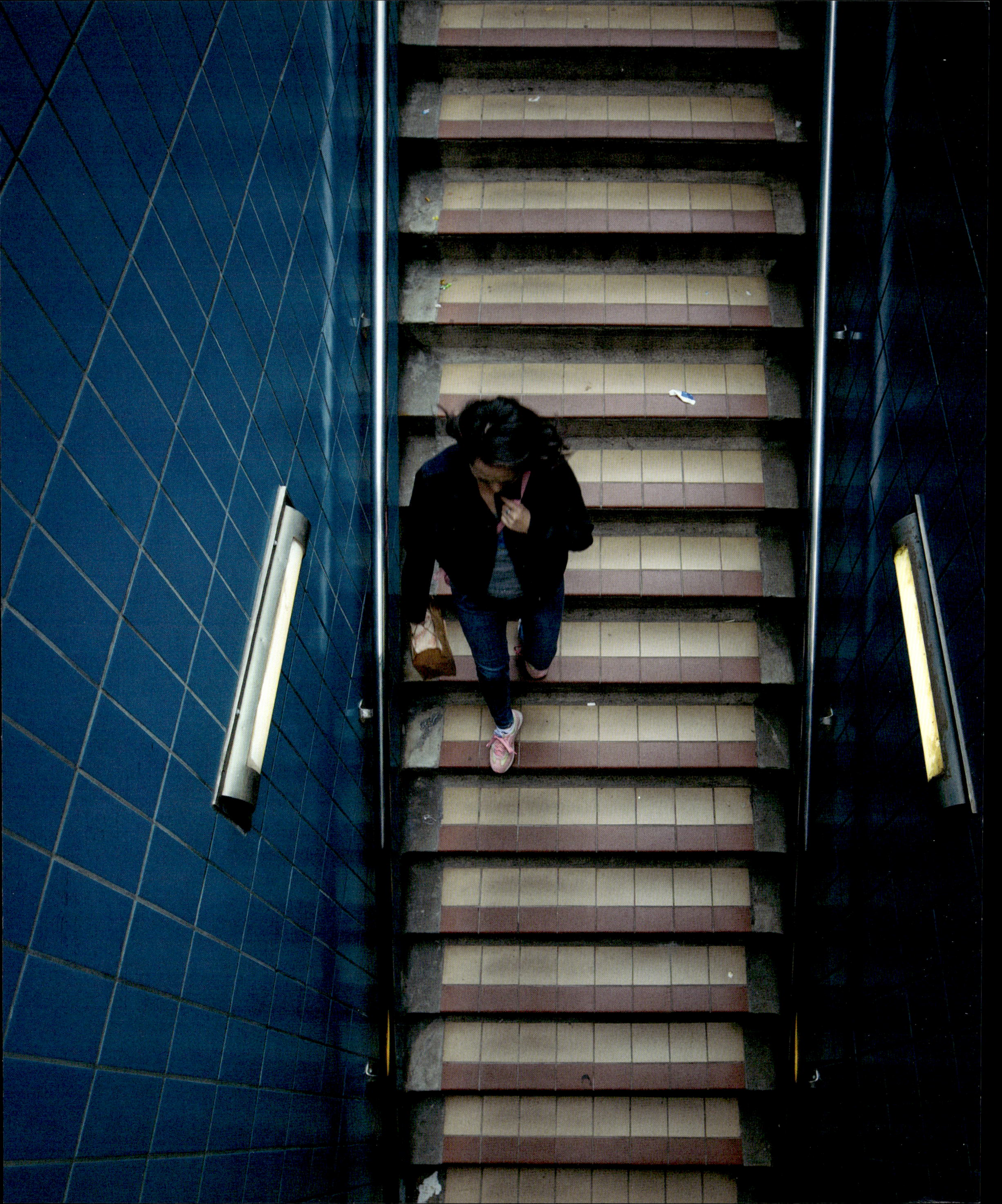

SAFETY'S BEST KEPT SECRET

JEAN-CHRISTOPHE LE COZE

In his novel *The Purloined Letter,* Edgar Allan Poe writes a story that reveals that things that are obvious – *right in front of our eyes* – are sometimes invisible. In this story, the police in charge of retrieving a stolen letter of great importance know that this letter is hidden in the robber's flat. However, the police fail to find the precious piece of paper. Poe's imaginary detective, Dupin, finds the letter by imagining the robber's mind. The robber's intention is indeed to outsmart the police who, he rightly thinks, will likely look for the letter in hidden places. The letter is, in fact, in sight on the desk, though it doesn't match the description. The letter was there all along, but no one had thought that it would be plainly visible. At one level of interpretation, this story is about missing things when they are right in front of us. The same could be said about strategy in safety.

Safety is the product of how a multitude of artefacts, people and institutions interact in digitized and globalized contexts. So, it's daunting to single out just one idea with practical leverage. Yet, it is worth thinking of safety as *strategy*. BP suffered a series of dramatic disasters within a few years (Texas City, 2005, Prudhoe Bay, 2006, Deep Water Horizon, 2010) all linked to *strategic* choices by top management (e.g. strong cost cutting policies, removal of safety centralized expertise). It's one case study which has become a strong warning against any ambitious strategies which would sacrifice safety for production. Adjusting companies' strategies in tough markets is a challenge, but when organizational limits are reached, serious events happen. These limits are experienced by workers, who *must be heard* at companies' highest levels as strategies unfold. What happens at sites needs to be heard at headquarters.

Remaining within these limits is far from an easy task. Limits are not always easily perceived by top management: people can compensate, absorb and cope with ambitious strategies, and it's not in their interest to admit if they can't compensate, absorb or cope. Moreover, in a competitive world, limits exist to be pushed. This simple idea, which no one with hands-on experience would likely deny, is unfortunately never part of organization's safety programs.

This safety narrative, which includes questioning top management decision-making and its influence on daily practices, is not mainstream in safety research and practice. It is indeed easier to consider that other people have failed to compensate, absorb or cope with operational challenges in tough business environments than to admit a flawed strategy. In fact, successful companies in safety do the opposite; they correlate safety events and workers' and managers' practices to strategic contexts, and adapt their strategies accordingly. But most companies are quiet about it. Safety is then considered mostly as a problem of front-line workers and their mistakes, not of top management and their strategic errors. This is safety's best kept secret.

Dr Jean-Christophe Le Coze is a safety researcher at the French national institute for environmental safety, INERIS. His activities combine ethnographic studies and action research in safety-critical systems. He's the editor of *Safety Science Research: Evolution, Challenges & New Directions* and the author of *Post Normal Accident: Revisiting Perrow's Classic.*

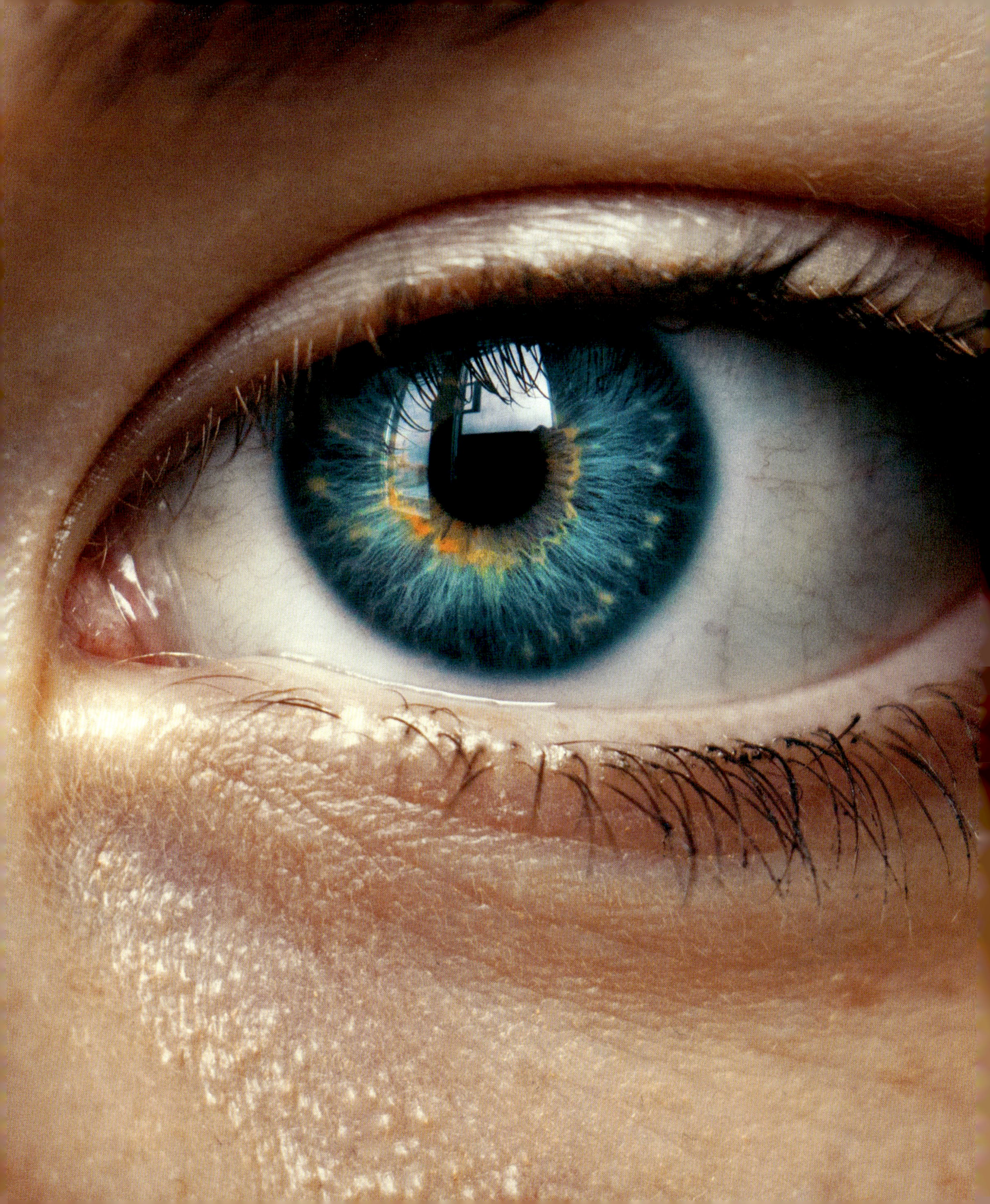

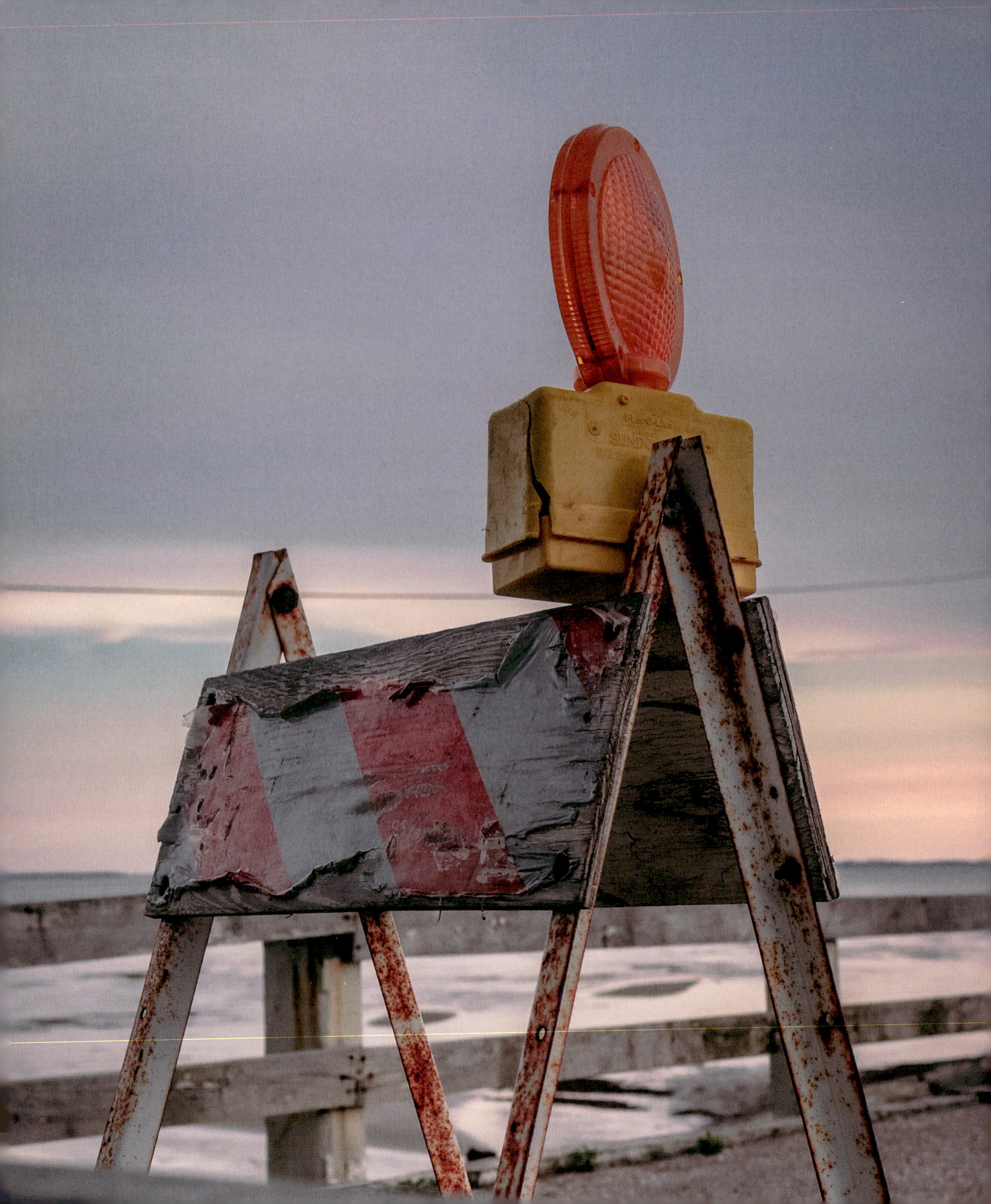

IF YOU'RE RISK ASSESSING A TASK, UNDERSTAND THE TASK FIRST

BRIDGET LEATHLEY

People often misunderstand and miscommunicate what they are risk assessing. If the task is misunderstood, significant hazards can be missed, and the wrong controls proposed. Accidents will happen. People will be hurt.

When supporting a facilities team with risk management, I ran a meeting to review their risk assessment for "changing a fuse." My first question was simple: "What sort of fuse are you changing?" "I guess it's the fuse in a plug," said one person.

"No, I think it means changing the circuit breaker in the distribution board," suggested another. A third person looked confused: "I don't think it would be either of those. It's the fuses in the transformers". The risk assessment was vague enough to apply to any of these, and general enough to be of use to none.

If you are risk assessing a task, start with a task analysis. For a simple activity, the task analysis can be a list of the steps needed, on the back of an envelope if necessary, or in a notes app on your phone. For a more complex project or operation, a hierarchical task analysis shows how each task goal can be described in more and more detail.

If you base your hazard identification on *your* idea of what people do when they change a fuse, rather than what they *actually* do, you will miss something. Don't sit alone in a room writing a task analysis. The best task analyses are produced by a team of people: those doing the task, those managing the task, and where practical, those impacted by the task.

Although a task analysis sounds like an extra, time-consuming step in the process of risk assessment, it is an excellent way of getting people talking about the task you are assessing. Creating a task analysis for changing a fuse (once we agreed which type of fuse change we were assessing) revealed other misunderstandings about who could carry out the task and the controls needed.

On a project in an office setting, a task analysis quickly identified there was no safe way to carry out work at a height over a desk. Moving the desks to use a ladder would involve heavy manual lifting and disruption to electrical cables. A technician admitted to waiting for people to go home and climbing on the desks, working alone. We purchased specialist equipment for working over desks safely, cutting down the need for overtime payments as well as reducing the risk of a technician falling when working alone.

Task analysis does add time upfront to a project, but you will make that time up later, because a shared understanding of the task means you will select the most effective controls to manage risk.

Bridget Leathley, CMIOSH, AFBPsS, started her career in Human Factors in IT, before applying her skills in high-hazard industries. She migrated to OSH but retains a passion for helping organizations to use technology to make workplaces safer and healthier. Alongside consultancy and training, Leathley writes blogs and articles.

WHY HR & LEADERS NEED MORE COURAGE

DAVID LIDDLE

The management of conflict, and particularly bullying and harassment, has perplexed Human Resources professionals, line managers and organizational leaders for many years.

Much like coronavirus, unresolved conflict at work can be an invisible killer. For those involved, directly or indirectly, conflict can generate untold amounts of fear, stress, isolation and anxiety. High profile cases have demonstrated that an employee's mental health is affected, and in the most serious cases, people have been known to take their own lives. In the UK's National Health Service, increasing levels of data suggest direct relationships between incivility in the workplace and adverse patient outcomes.[1]

The paradox is that the policies designed to create psychological safety at work make people much less safe. The traditional policies and procedures offer a blunt instrument for managing conflict at work. They are reductive, believing that there must be right or wrong and a winner and a loser in every case. They provide a mirage of justice and an illusion of fairness.

The reality of what I call the GBH processes (grievance, bullying, harassment), plus traditional disciplinary procedures, is that they perpetuate a negative, damaging, and corrosive tone within workplaces. They undermine trust, infantilize the workforce, sow the seeds of division, impede creativity and hurt people. The paradox that haunts many an HR professional is that the very policies designed to resolve workplace issues make them a lot worse. It is like walking up to an individual exhibiting signs of stress and distress, pouring a bucket of cortisol over their heads, and then yelling at them for not being more rational. This is not a great way to resolve a problem at work.

This crisis has shown us that people can be trusted to get on with their jobs, are loyal to their employers and show flexibility, resilience and dedication, if they are trusted and given the freedom to flourish. Companies should simplify their rules and processes so that they align to three simple principles:

• Do the right thing.
• Follow our values.
• Operate within the law.

Organizations must develop and agree on behavioural frameworks which are aligned to their core values. These frameworks should clarify the aligned and the misaligned behaviours that they expect (or don't expect) from their managers and employees. Employees, managers and others should then be equipped with the skills, the support and the resources that they need to hold themselves and each other to account.

Managers and leaders should be trained in the vital skills they need to excel in the new normal – positive psychology, nudge theory, principled negotiation and nonviolent communication, to name a few.

Finally, HR, unions and leaders should collaborate to reframe their divisive GBH policies. The entire process should be repurposed, with emphasis on early resolution between the parties, supported by processes, such as restorative conversations, coaching and mentoring.

Whatever the new normal holds, this has got to be better for the wellbeing of our employees, customers and businesses.

1. https://qualitysafety.bmj.com/content/28/9/750.full

David Liddle is Chief Executive of The TCM Group. TCM is an award-winning provider of mediation and investigation services, employee relations consultancy, cultural transformation and leadership development programmes. Liddle's latest book, *Managing Conflict* (Kogan Page/CIPD) is available now.

BE BRAVE!

FOCUS ON OSH EDUCATION FOR YOUNG EMPLOYEES

MARIA LINDHOLM

Young employees, aged between 15 and 24 years, are at higher risk for occupational accidents and injuries. Even though they have less serious and less frequent musculoskeletal disorders than older employees do, there is an issue of early exposure that will become more serious later in their working careers[1].

I have studied[2] university students' perceptions about occupational safety and health (OSH). For example, their responses to the open-ended question: "Tell in your own words what OSH means" indicated that a third of the participating students had a fairly stereotypical view of OSH. Regarding students' impressions of their safety maturity level in relation to the Bradley Curve, slightly more than one third of the students chose the two first stages (reactive and associated with higher injury rates) of the Bradley Curve and almost two thirds chose the two more developed stages (proactive and associated with lower injury rates)[3]. In addition, a third believed that OSH slows work down, and almost half slightly agreed or agreed that they take risks when in a hurry. These results indicate that the maturity level of OSH knowledge varies among students and that some of the students' maturity level is poor. It could be asked whether these perceptions and attitudes will be reflected in their future careers.

It is essential to prevent young employees' occupational accidents and injuries to ensure long careers and the opportunity to retire healthy. OSH education and training is considered important in the early stages of working careers, but the amount and quality of OSH training in the first year of a new job varies considerably[4,5] indicating that young workers may be especially vulnerable in this industry. In Norway, it is possible to enter the construction industry as a full time worker at the age of 18. The aim of this study was to explore how young construction workers are received at their workplace with regards to OHS-training. The study was designed as a qualitative case study. Each case consisted of a young worker or apprentice (< 25 years. As Schulte et al.[6] noted, schools may be the only setting in which some employees receive adequate OSH information.

I highly recommend that all educational levels focus on OSH teaching and that employers pay special attention to the orientation of new, young employees in order to keep young employees safe and healthy.

1 Laberge M, Ledoux E. Occupational health and safety issues affecting young workers: A literature review. *Work*. 2011;39(3):215-232. doi:10.3233/WOR-2011-1170
2. Lindholm M, Väyrynen S, Reiman A. Findings and Views on Occupational Safety and Health Teaching at Universities. *Work*. 2019;64(4):685 – 695. doi:10.3233/WOR-193030
3. DuPont. The DuPont Bradley Curve. 2018. http://www.dupont.com.au/products-and-services/consulting-services-process-technologies/brands/sustainable-solutions/sub-brands/operational-risk-management/uses-and-applications/bradley-curve.html.
4. Holte KA, Kjestveit K. Young workers in the construction industry and initial OSH-training when entering work life. *Work*. 2012;41(Supplement 1):4137-4141. doi:10.3233/WOR-2012-0709-4137
5. Smith PM, Mustard CA. How many employees receive safety training during their first year of a new job? *Inj Prev*. 2007;13(1):37-41. doi:10.1136/ip.2006.013839
6. Schulte PA, Stephenson CM, Okun AH, Palassis J, Biddle E. Integrating occupational safety and health information into vocational and technical education and other workforce preparation programs. *Am J Public Health*. 2005;95(3):404-411. doi:10.2105/AJPH.2004.047241

Maria Lindholm is a doctoral candidate at the University of Oulu, Finland. Her research interests are occupational safety and well-being at work. She completed her MSc in Industrial Engineering and Management in 2016 and performed a research exchange period in Karolinska Institutet (Sweden) during 2019.

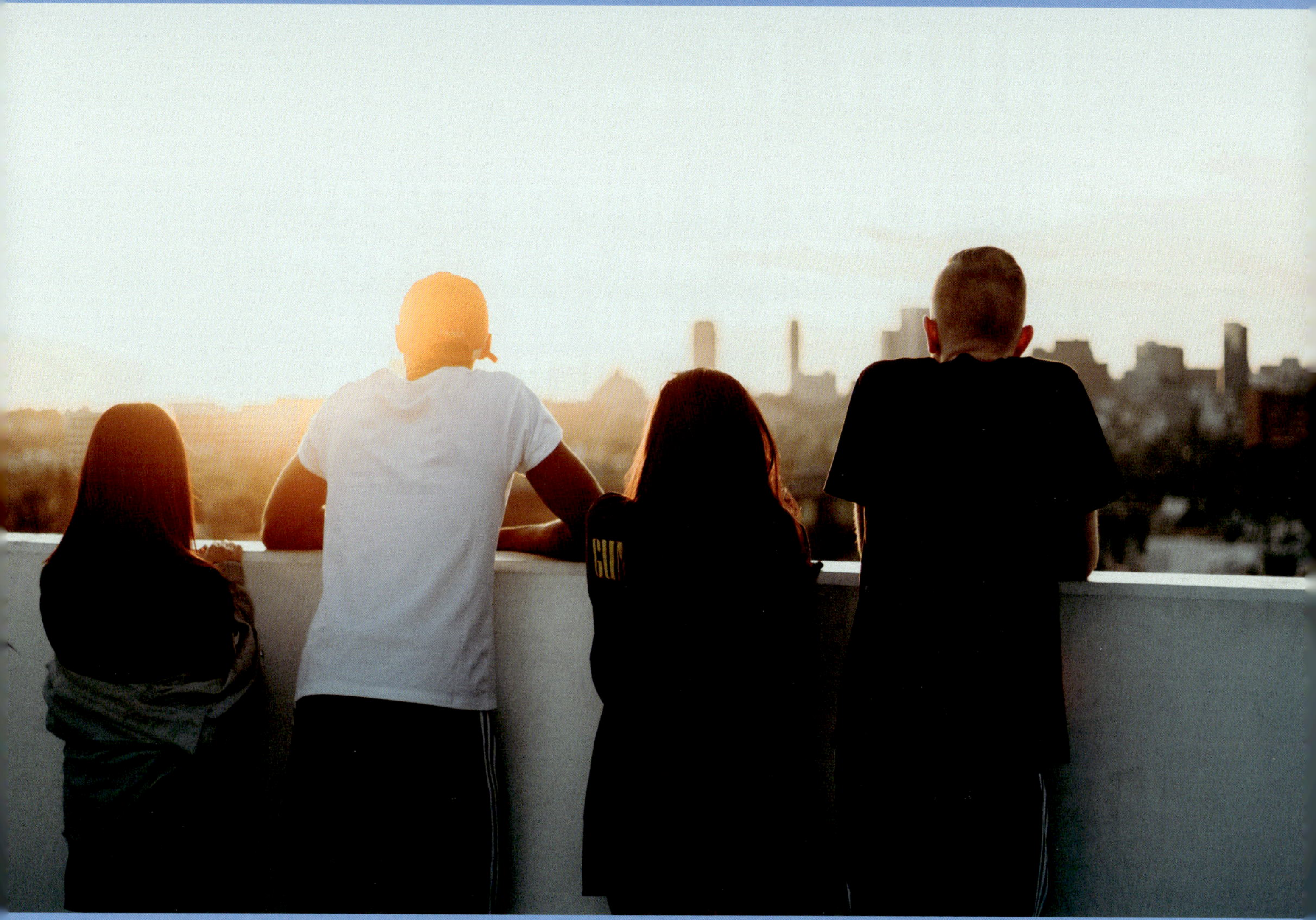

WHAT IF
DR NATALIE LOTZMANN

WHAT IF...

THERE WAS AN APP ON EVERY SMARTPHONE WHERE WORKERS COULD RATE THEIR HEALTH & WELL-BEING AT WORK, THEIR STRESS AND SATISFACTION, THEIR WORKING CONDITIONS AND THE LEADERSHIP BEHAVIOR IN REAL TIME...

...AND THE DATA WOULD BE TRANSPARENT TO EVERYONE?

WHAT IF...

WE WOULD NOT ONLY CARE FOR FINANCIAL REPORTING BUT MAKE THIS "HUMANCIAL REPORTING" MANDATORY IN THE ENTIRE SUPPLY CHAIN?

Dr Natalie Lotzmann is Chief Medical Officer at SAP. She develops concepts, programs and innovative KPIs focusing on a healthy culture. Her work has been recognized with numerous awards for "Health", "Diversity" and "Equality", including several Move Europe Awards and Corporate Health Awards.

TIME FOR SAFETY TO GET ENGAGED?

JIM LOUD

'Engagement' has become a recurring buzzword in safety discussions and literature. There is a growing consensus regarding its importance, not just to safety, but to any organizational objective. Despite considerable lip service promoting an engaged workforce, global results from Gallup show that 85% of employees are disengaged from their work. Gallup's *State of the Global Workplace* (2017) put the problem like this: "They (the workers) are indifferent to your organization. They give you their time, but not their best effort nor their best ideas. They likely come to work wanting to make a difference – but nobody has ever asked them to use their strengths to make the organization better."

The disengagement problem is far bigger than safety, but it doesn't have to be that way. Workers who feel like business partners and stakeholders in *their* company naturally want their actions, and the actions of others, to help their company succeed. Transforming disaffected workers into work *owners* and champions for continuous improvement is, of course, a formidable challenge and won't happen overnight. Those of us in safety can, however, move the needle by promoting treatment of workers as assets and problem solvers rather than liabilities in need of fixing via behavioral modification and ever more command and control. This will require a fundamental shift in thinking for many.

My experience has convinced me that, given the right environment, employees will readily help find ways to improve how their work gets done – including how safely it gets done. But attempting to change employee behavior without changing the systems and environment that impact the work is an exercise in futility. You can't train in, preach in or enforce in engagement. If employee involvement is really what you want, you'll need to give up some of your perceived control and share it with your employees. Listening respectfully and humbly to what the workers have to say is a good first step, but you (and your company) must also respond to what you learn from the workforce in a timely and serious manner. Failure to do so is the surest way to lose worker trust and productive cooperation.

Unfortunately, many in the safety practice remain wedded to traditional command and control tactics. Others seem genuinely threatened by more worker safety participation on their perceived turf. Sharing longstanding safety responsibilities with those most directly impacted will require, for many of us, a new way of looking at safety and a willingness to break from the status quo. It will also require courage and integrity to do the right thing. Despite my impatience, I firmly believe the vast majority of safety professionals want what is best for safety. But change is always difficult, and often damnably slow. Still, I optimistically believe that as the advantages of an engaged workforce become ever more obvious, the safety profession will, in fact, help lead their companies in transforming themselves into more satisfying, productive and safer places to work.

James Loud, MPH, MS, CSP, is a safety consultant with over 40 years' experience in safety and management. He is a frequent speaker at national and international conferences, webinars and university classrooms and has authored numerous articles and papers on safety and safety management.

A NEW LENS ON CULTURE

JOAN LURIE

Culture emerges from the interactions between the parts of the organizational system.

'X' Company was in trouble! Their margins were declining, and sales were being lost to their main competitor, who had changed their business model to become a digital company. They had responded by commissioning an online platform; however, this still was not active 18 months later, and their culture was eroding. Something was going on, which was getting in the way of them building the technology solution, but the problem wasn't just technical.

A burning platform isn't always enough

They knew they had to change but they were 'stuck'; each department blaming the other. The Marketing function, for example, had agreed to their new strategy versus service role – including no longer responding to sales' information queries now available online, but they were still spending 20% of their time addressing these.

The CEO and the Marketing leadership team were frustrated: "We need to put out another round of comms and make it a directive now". But that was a solution based on the wrong assumption. When they got on the balcony to "see their whole system", they uncovered their hidden agreements and competing commitments. The problem wasn't with their formal communications or changing Sales. The Marketing team was equally part of the problem.

All behaviour is circular (not linear)

Looking through the systems lens made visible their patterns of relating and their co-creation of this 'unwanted' outcome. Explicitly, Marketing had agreed their new role with Sales and had 'clearly' communicated the alternative source of information, but implicitly, they were giving a contradictory message by not setting boundaries and not re-directing the queries. The entire Marketing team had an unspoken agreement to not be the 'baddies' and say no to Sales, thereby also messaging Sales: "It's okay to continue".

Uncovering the 'hidden' implicit feedback loops which keep systems stuck

With good intentions to 'help', they were keeping the system in this perfectly co-created circular feedback loop. Sales got their answers and in turn, Marketing received positive feedback for being the 'helpers'. They all had a 'hidden' systemic contract to not change. Marketing realized they had to give up their role of being the 'helpers' to change the pattern rather than blaming Sales.

Change the role, change the pattern, change the culture

The leaders now understood that instead of trying to change others – individuals' values and behaviours – they had to uncover and change their competing commitments, implicit 'agreements', and circular patterns, which they did.

Within a few months, applying this systemic approach on a company-wide scale, 'X' Company became 'unstuck' – a new culture was emerging, their digital platform working – the margin erosion stabilized. The business was turning around.

We are too used to paying attention to technical solutions and individual and group behaviour, rather than changing roles, relational patterns and rules of engagement between the parts of our systems. Once we reframe and understand organizations through this systems lens, everything changes quickly. It's time to rethink culture in organizations.

Joan Lurie is founder and CEO of Orgonomix. A developmental psychologist and organizational change consultant, she applies her unique systemic methodology to support leaders to grow their adaptive capacity, emerging more fit for complexity, transforming themselves and their organizations' culture and operating models.

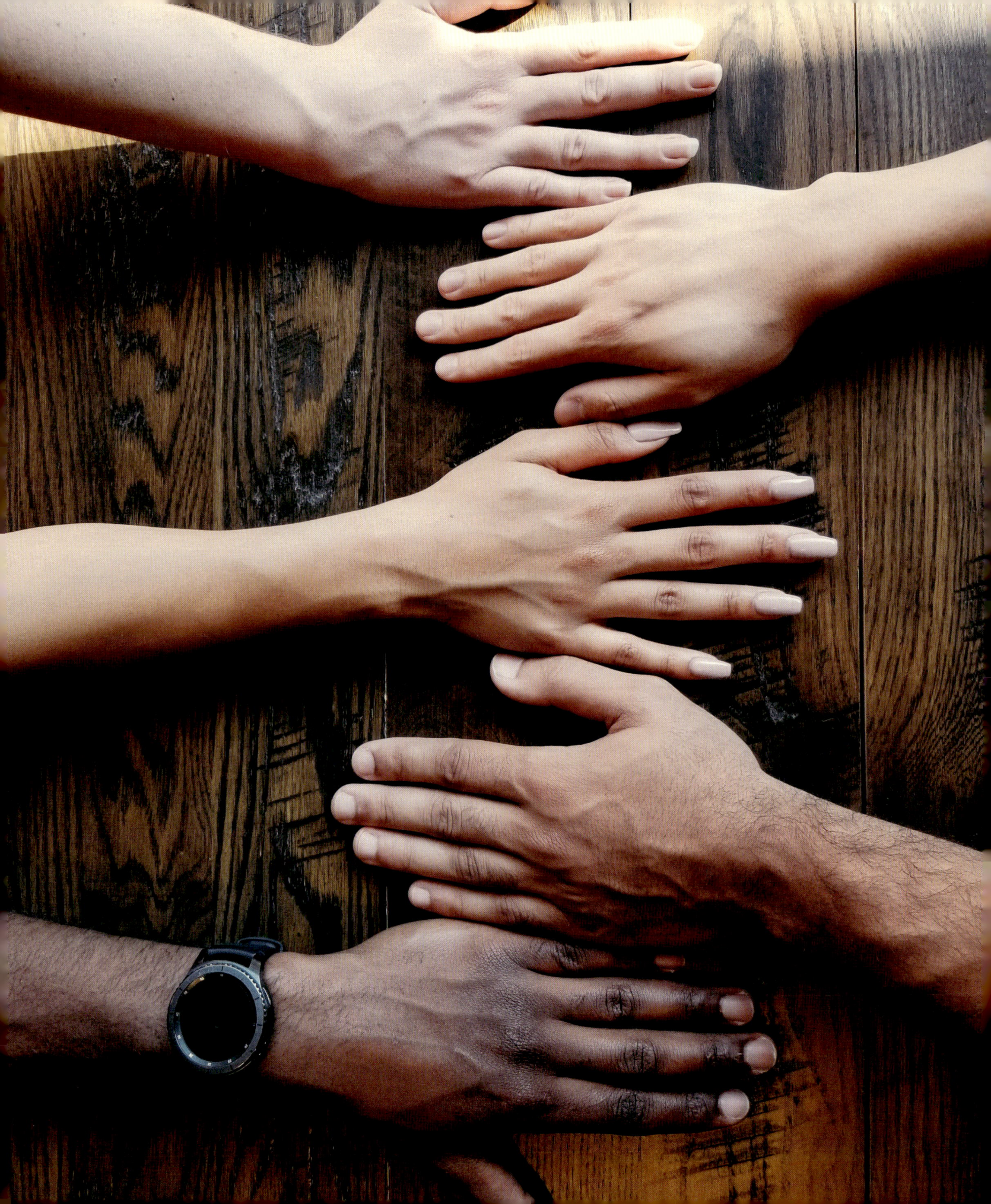

COLLABORATION IS KEY

DAVID MAGEE

There is a cliché used in OSH: "Any organization is only as strong as its weakest link."

Research shows that poor communications and human factors play a role in a significant percentage of OSH incidents. Particularly at-risk employees are those within their first twelve months of employment. Communication is a two-way process. It requires all interlocutors to have the requisite levels of subject knowledge and subject language to be successful.

OSH uses a unique and complex communication system comprised of different colours, shapes, symbols and terminologies. These have been developed and agreed upon by international bodies such as the ILO, the ISO and UN. They are universally accepted and used. This communication system can become even more specialized depending on different industry sectors. Examples can also be seen at home on everyday domestic products and appliances. In public spaces too, for: fire, evacuation, electrical and construction hazards, access and egress, traffic and medical aid. There are hard and soft OSH literacy skills, such as being able to use safety equipment, identify and deal with hazards, communication skills and working safely with others. OSH literacy abilities can be taught and measured from entry to an advanced level, and it involves a continuum of lifelong learning. Indeed, OSH literacy meets all the requirements to be classed as a 'literacy' comparable to the many others in this 'information age' of 'multi-literacies'. However, this has not happened yet.

As a result, globally, most young people leave school and enter in to training, employment and independent living without the requisite OSH literacy skills they are likely to encounter or need. In addition to a lack of education, there are other barriers which can restrict a person's ability to engage and comply with OSH information and training. These can be cognitive, physical and linguistic, especially with English as the lingua-franca of the international workplace and OSH. OSH uses a lot of terminology not taught as part of any native or English language programs. There are also age, gender, economic, ethnic, social, cultural and even religious barriers.

Legislation to tackle this issue, such as the UN Convention on Human Rights and EU directives, put a burden of responsibility onto employers to ensure that their OSH information and training is accessible to all. However, many employers are unaware to the extent of learning issues which may exist within their workforce. Nor do many have the time, training or resources needed to adapt their language and materials to make them accessible to all employees. For years, employers have been asking educators and other stakeholders to give future workers skills more aligned to their needs.

There is obviously a break in the transitional communications chain. We need more collaboration between all stakeholders to develop a robust, comprehensive, and cohesive system to assist young people to safely and successfully transition from education to employment and training. Instead of the traditional model of cascading information down, we also need to build OSH literacy skills from the bottom up.

J. David Magee is the founder of OSH literacy.org, an NGO which advocates for the recognition and teaching of OSH literacy as an essential Key Skill literacy.

TIME FOR A NEW PARADIGM?

THIERRY MEYER

The world we live in is increasingly interdependent and interconnected. But at the same time individualism and individual performance is often held up as a model for success. Should we continue with this vision? Major negative events such as the crises we have recently experienced reveal that these models very quickly reach their limits and that we need to reconsider our way of thinking and acting. It is common to hear that there will be a 'before' and 'after' Covid-19, but what are we really going to learn and put into practice? Shouldn't we rework our thought patterns in terms of essential priorities and comfort? For example, was the reason that many people rushed to buy toilet paper a key priority or was it comfort? Was the thinking rational or emotional? Did conventional risk management models anticipate this? Did the official messages from the authorities and industrialists to appease the population have the expected effects?

Isn't it time to try a different paradigm other than one based on statistics and probabilities? Moreover, what do such probabilities mean for the uninitiated? Are they just numbers that no-one really understands or that are difficult to explain in terms that make sense to everyone.

The classical risk model consisting of likelihood of occurrence and severity seems no longer suitable for today's situations. Nowadays, we observe that *'black swans'* are occurring more frequently. Why continue with this likelihood of occurrence? Wouldn't it be more reasonable to focus on high-potential severity scenarios, whether for humans, the environment or infrastructure? Believing that a low probability will protect us is a concept to be abandoned, would preparing for severity be mitigated be the short-term solution? But this must be accompanied by the indispensable integration of human factors as well as the duality of essential versus comfort priorities.

Let me rephrase the objective raised by Andrew Sharman with *One Percent Safer* as *"How can we make our family, friends, neighbors, colleagues, coworkers and fellow citizens more aware of the dangers and keep them out of harm's way?"*

We should focus on providing simple and effective strategies for safe living and working. Make it simple enough that it could be explained to grandparents or young kids. I would suggest being very pragmatic and avoiding a scientific or technocratic style. To change anything, you need to understand.

- Listen to the needs and priorities (essential or comfort?): **LISTEN** and **CARE**
- What is the intensity of the potential harm?: **UNDERSTAND, THINK**
- Is it possible to mitigate the threats?: **EVALUATE**
- Can I afford to implement the mitigation? if not, could I sustain the harm?: **ACT, BE PREPARED**
- Who could help/support me so we are stronger together?: **SHARE, COLLABORATE**

I conclude with a quote from Maya Angelou, the famous American poet:

"Hope for the best, be prepared for the worst, and unsurprised by anything in between."

Dr Thierry Meyer heads the Safety Competence Center and a research group at EPFL in Switzerland. His former professional affiliations include several chemical industrial companies. His research interests focus on risk and safety management, safety culture, hazards in research laboratories, and decision making.

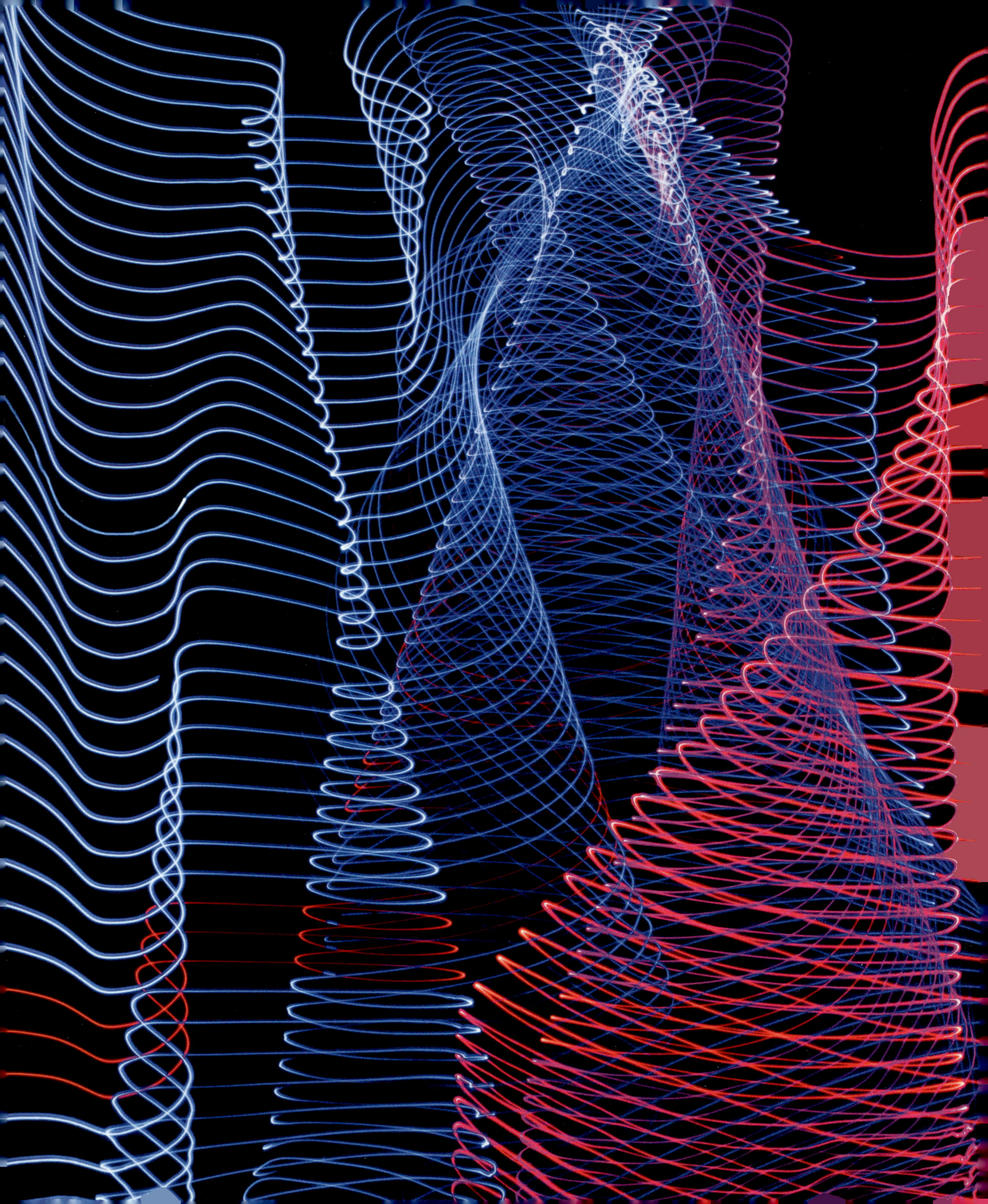

TAKE TIME TO BE HEARD

KAREN MCDONNELL

Take time to reflect, as 2020 has shown, situations can rapidly evolve and perhaps never return to the previous 'normality'.

Safety is a 'whole person, whole life' subject.

Lives free from serious accidental injury are achieved through exchanging life-enhancing skills and knowledge improving personal and organizational sustainability.

It has been said that "Every next level of your life will demand a different version of you". It is a good time to pause and prioritize not just your people, but the 'different version of you' they may require.

And when you have paused, remember the words of Maya Angelou: "If you want what you are saying heard, then take your time so the listener will hear it".

Karen McDonnell is OHS Policy Adviser, Royal Society for the Prevention of Accidents and a Past President of IOSH. McDonnell has a broad range of strategic and operational experience which is applied to influence and motivate others to think out with traditional boundaries towards sustainable working lives across the world.

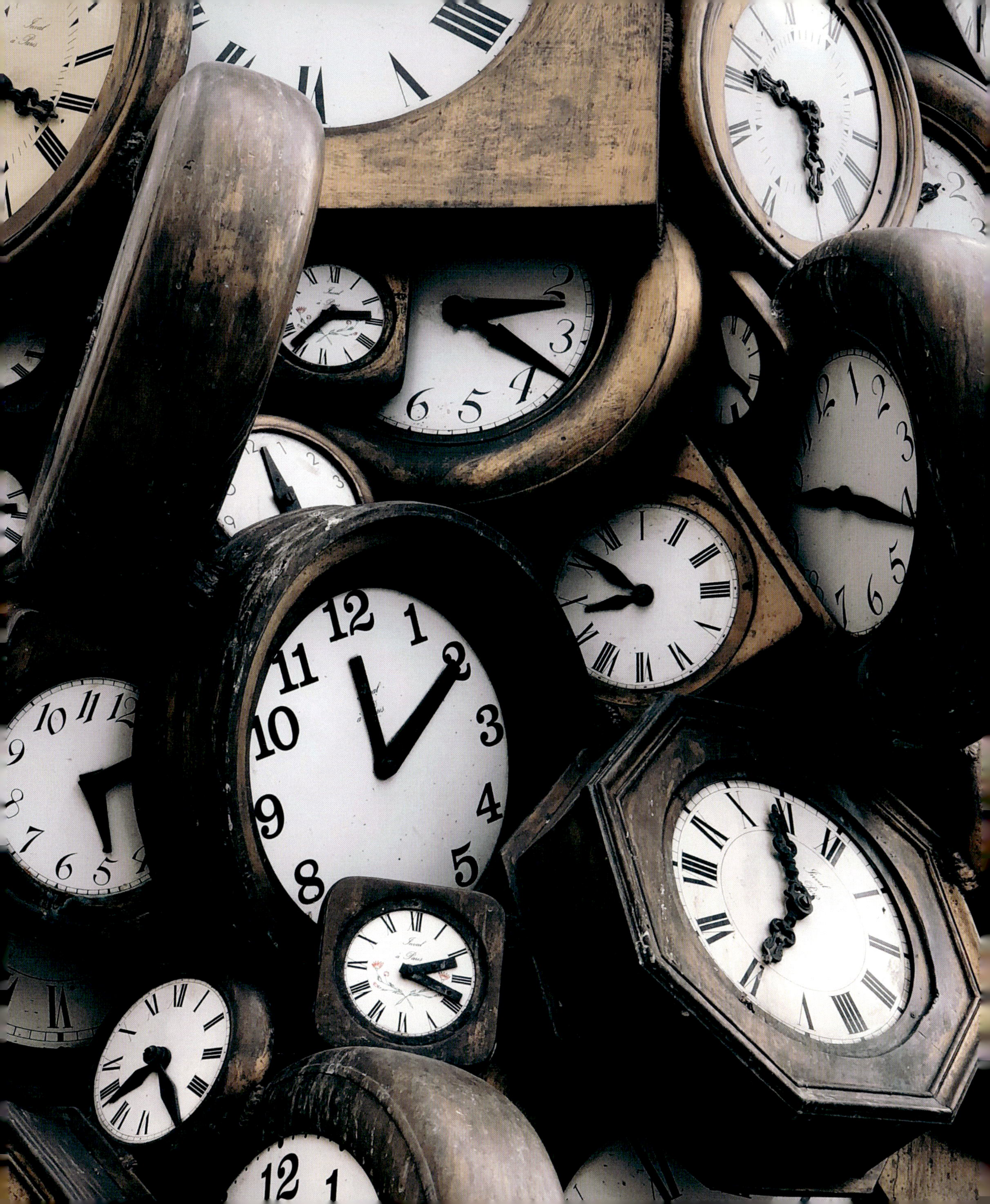

blink
blink
WE SUPPORT
OUR TROOPS
CYCLE
33
F.D
N.Y.
FIRE DEPARTMENT
CITY OF
NEW YORK

THE FUTURE OF HUMAN AND ORGANIZATIONAL FACTORS

KATHRYN MEARNS

Human and organizational factors (HOF), regulatory shortcomings and 'cultural' issues have been consistently identified as the underlying causes of major accidents in a wide range of industries. Recent examples include the 2008 financial crash, Deepwater Horizon, Fukushima Daiichi and the collapse of the Carillion consortium in the UK. There is evidence that although industry and regulators claim to 'learn lessons' from major accidents, there seems to be an inability to apply the HOF lessons learned. The question is why? Is it because senior managers fail to understand HOF and its impact on safety performance? Is it because regulators are not informed or confident enough to challenge on HOF issues? Is it because 'culture' is too difficult and does not necessarily lend itself to conventional management interventions? These questions need to be addressed in order to achieve more resilience and reliability in organizations that play a critical role in maintaining society's safety, security and welfare.

The challenges faced by complex (multiple system components) and complicated (high level of intricacy) organizations, should not be underestimated, but their successful performance depends on a combination of positive and prudent leadership, and a competent, well-informed regulator. The workforce also plays an important role through reporting issues and providing innovative and practical ideas to make improvements, for example via safety culture/climate assessments. These assessments allow the workforce to provide information about the issues that challenge them on a day-to-day basis, e.g. perceived lack of senior management commitment to safety; inadequate communication (too much of the wrong sort or too little of the right sort); inadequate procedures (badly written, out-of-date or failing to cover how work is actually done); fear of raising challenges about safety and lack of organizational learning. Often, such assessments and surveys tend to lack context, so I suggest that the assessment process should include discussion groups or interviews for a more coherent picture of the circumstances in which the issues occur and the solutions proposed to address them.

The way forward is for regulators and senior managers to be better educated about the importance of HOF in maintaining safety, security, welfare and performance and to understand the value of workforce involvement in this process. Comprehensive climate/culture assessments provide critical intelligence for managers and regulators but the data must be used appropriately. If senior management do not respond to the results of these assessments their credibility will be eroded, for example if proposed solutions to problems cannot be implemented, then leaders should be willing to explain why. Leaders need to understand HOF issues in order to *apply* the lessons learned, rather than simply claiming they have 'learned lessons', without implementing follow-up actions. As public inquiries into major accidents have demonstrated, HOF issues can reflect the underlying conditions that demonstrate the organization is drifting towards failure – and that failure can sometimes end in disaster.

Kathryn Mearns is a Human and Organisational Factors Specialist with over 25 years experience as an academic, regulator and practitioner in a range of safety critical industries. She currently works for Jacobs as a HOF Specialist in the nuclear industry.

PUBLICIZE WORKPLACE INJURY RATES

DAVID MICHAELS

Many employers have active safety and health management systems and strive to prevent all injuries and illnesses. Some, however, need additional incentives. Transparency can be a powerful driver of behavior. Making workplace-specific injury data available to the public could 'nudge' more dangerous employers to better protect their workers.

Why? First, employers compete to attract the best possible workers at prevailing wage rates. Although workers can generally learn about wages and benefits at prospective employers, information on safety is harder to come by. This is a problem because there is tremendous variation in injury rates among employers, even in the same industries in the same areas.

In the US, hospitals and nursing homes are among the most dangerous places to work, with injury rates higher than construction or coal mining. The chances of being hurt in one nursing home can be five times that of another facility in the same town. Just as consumers benefit from information regarding which cars have the better safety records, workers would benefit from ready access to information on injury risks in making job choices. Nursing homes with low injury rates would become more attractive to workers while those with high rates would face pressures to improve.

Injury rate transparency can work through a second path. Evidence shows that firms that focus on quality production generally have low injury rates because work processes are tightly managed. High injury rates can indicate poor management and lax standards. If consumers or other businesses care about product or service quality, injury rate disclosure can be a proxy of operational quality. It's not surprising then that many responsible employers, proud of their low injury rates, support safety transparency.

Returning to the nursing home case: high worker injury rates may reflect inadequate staffing or lack of investment in safety equipment like lifts to help patients get out of bed without injuring the worker or dropping the patient. If their worker injury rates were public, more dangerous nursing homes would face pressure to improve safety performance, not only to draw skilled job seekers, but to attract potential patients.

Research demonstrates that carefully-crafted transparency policies can improve public safety. One compelling illustration of this is posting health inspection grades in restaurant windows. Consumers, eager to avoid food-borne illness, take these grades into account when deciding where to dine. After a restaurant grading program started in Los Angeles, revenues rose at establishments with high marks for food safety and fell at those with low ratings. More importantly, hospitalizations for food-borne illnesses decreased significantly.

Making injury rates public would also enable employers to benchmark their safety program with other employers in the same industry, facilitate research into factors that increase injury risk, and enable improved evaluations of injury prevention initiatives.

One caveat: for this approach to work successfully, workers must be encouraged to report injuries. In order to appear safer, some employers may attempt to reduce reporting rather than prevent injuries. This must be strictly prohibited and accompanied with appropriate penalties if it occurs.

Dr. David Michaels is an epidemiologist and Professor at the Milken Institute School of Public Health of George Washington University. He is the longest serving Assistant Secretary of Labor for the Occupational Safety and Health Administration (2009-2017) in the agency's history, and was Assistant Secretary of Energy for Environment, Safety and Health (1998-2001).

POSITIVE CHANGES FROM THE PANDEMIC

LUIZ MONTENEGRO

I believe that the COVID-19 pandemic brings us a number of valuable lessons to get a bit safer every day that we must take forward.

Despite the reduction in volumes produced, there was plenty of activity going on at Carlsberg's sites during the past few months. Nevertheless, we were able to observe a surprisingly strong decrease in the number of work-related injuries. While reduced volumes and workload should imply in a reduction of exposure to risk, we wondered why the reduction in the incident rates observed was far more drastic than the reduction in business activities.

What did we find?

During this crisis, there was a much higher level of awareness and adherence to rules and standards, a more independent operating crew working with less supervision, and an overall supportive attitude towards each other. The seriousness of this disease and the need for drastic control measures at the public level resulted in a spirit of solidarity and raised the importance of safety and health in all spheres of our lives.

In other words, everybody quickly understood *why* the control measures were put in place and why it was so important to follow and to help others to conform.

If we extend this learning to the other dimensions of health and safety, it highlights the importance of focusing our efforts on the prevention of severe incidents, because it provides clear and appealing purpose, making it easier to communicate and engage people for the cause. At Carlsberg, the focus on our "Five Life-Saving Rules" over the past two years has not only eliminated work-related fatalities, but also brought down the accident rates. Pursuing the most important zero – zero fatalities – was far more important and appealing to all levels of the organization, and the result has been an enhanced culture of prevention that permeates to the other things we do. The COVID-19 pandemic came to reinforce and accelerate the change in culture we were already observing.

Times of crisis are always times of big opportunities. I believe we will all exit the pandemic much stronger in terms of risk awareness, compliance to fundamental rules, and above all, with a stronger level of engagement for health and safety. It's really a unique opportunity to change gears to a more empowered work force, more inspiring leadership and an accelerated development of self-managed teams. These changes could drive the continual improvement in health and safety, all based on a clear mutual purpose and on a truly caring attitude. Maybe the much-desired cultural stage of consistently caring for each other is just around the corner. We must look carefully into our current processes and systems in order to make them enablers of this transformed culture. The role of leadership during these times of difficult transformation is even more important. Together, we will be determined to the speed and the success of this change, and most importantly, maintain it.

Luiz Montenegro is VP Group Health & Safety at Carlsberg Group. A chemical engineer with post-graduate studies in Health & Safety Engineering and an MBA in Modern Business Management, Montenegro had 28 years' experience in the chemical and food and beverage industries in South America, Europe, Middle East and Africa before joining Carlsberg.

SOCIO-ECONOMIC PROFILING OF WORKERS

ALEX MORALES

As companies walk the talk of 'safety culture' and 'Vision Zero', they encounter different barriers in their efforts to reduce their accident and occupational disease indicators.

A common denominator in those who have been successful in maintaining a downward trend is the adoption of a systemic perspective that looks beyond the engineering and technical elements of risk control.

So-called 'soft skills' are among those new perspectives and the number of consultants that offer their 'proprietary' design and programs to introduce those competencies in the workplace has increased exponentially in recent years.

In our experience in Chile, introducing Ronald Heifetz´ Adaptive Leadership theory to Health and Safety management has been rewarding, and those concepts have contributed to increasing worker's participation in risk assessment and risk control, with greater ownership and under a new style of leadership.

In that context, a social worker that was part of the HR department of a Chilean manufacturer of electrical and metal-mechanical infrastructure decided to go a step further, and she called the company's safety manager to discuss a project to analyze the relationship between workers' socio-economic profiles and accident rates.

The team project found that an important number of highly-skilled workers that performed some of the critical tasks in their production process were in deep debt with multiple creditors, mainly through overusing credit cards that retailer and department stores offered them.

They also identify family psychosocial issues, especially drug use problems in teenage children, which were a major concern for many workers. When they presented the results of their project to the top management of the company, they obtained support for the implementation of new HR and Health and Safety practices.

These new practices encourage workers to seek advice from financial professionals from the company if they are in financial trouble, and the company even provided soft credit to help them sort out their debts. The new practices also encourage workers to self-identify with their supervisor if in any particular day the emotional burden of troubles (family, illness, conflict, debt) was overwhelming, and the company offered a paid day off to that worker.

As a result of these practices, not only there was a sharp decline in incident and accident rates, but their workplace climate improved.

Based on this case experience, my contribution to a safer world, especially in the new COVID-19 world that awaits us, is an invitation to go back to our success as homo sapiens based on collaboration and mutual support, and to expand those skills to our workplace, our families and our societies worldwide.

Alejandro "Alex" Morales did an MSc in Occupational Medicine (1991, Aberdeen University), worked as occupational health physician in Chile (1993 – 2005), held managerial posts in Mutual de Seguridad CChC (2006 -2016). Morales presently works at Universidad Catolica de Chile as medical director of the workers' occupational health unit.

WHAT WE SEE AFTER THE ACCIDENT

KEITH MORTON

I anticipate that I am in a minority of contributors to this book, in that I only become involved in cases after there has been an accident, injury or death. Over the past 25 years, I have advised hundreds of organizations and people who are under investigation or subject to prosecution by regulators, such as the Health and Safety Executive. By definition, in my line of work, sadly, safety has not been achieved. Is it possible to identify some causative factors which frequently recur?

The immediate cause of most accidents is generally easy to identify. What is striking to me, however, is that in almost all cases a series of individual acts and omissions coalesce in the moment of the accident. In isolation, each individual act or omission may appear insignificant. Even if foreseeable, in isolation it may not give rise to any significant risk. Organizations are fixed with liability because the law permits individual acts and omissions for which they are accountable to be aggregated. With the benefit of hindsight, the connections between each of them can be seen with clarity. Lawyer's shorthand often categorizes this as a "system failure". Yet accidents happen in the most sophisticated and well-run organizations with well-resourced and established systems for managing safety. In almost every case I have ever done, it is said, after the event, the risk assessment was not suitable and sufficient precisely because it did not address the very thing that has materialized.

Planners do not have the luxury of hindsight. Avoidance of accidents requires foresight. Risk assessments are a useful starting point. But they invariably assume everything will run smoothly; people will do as they are told; everything will be delivered on time and so on. That is often not the case. It is *never* the case when there is an accident. To be a blueprint for the safe implementation of work a risk assessment, method statement or plan should contemplate how the job might go wrong at the point of implementation. In my experience the following factors are key components to effective planning:

1. Anticipate human decency – most people want to do their best to help get the job done by going over and above what is expected of them, and in doing so can expose themselves and others to a risk of harm.

2. Anticipate human fallibility – rational people make mistakes and errors of judgment, and in doing so can expose themselves and others to a risk of harm.

3. Anticipate things will not go to plan – it is foreseeable that interactions between organizations and people at work will not go to plan, and this can expose people to a risk of harm. Ensure planning incorporates the agility to manage change.

All three considerations are commonly overlooked. Individually and collectively, effective consideration of these factors would have avoided very many of accidents, injuries and deaths that have given rise to the cases I have been engaged in.

Keith Morton QC is a leading legal practitioner in the fields of health & safety law, inquests and inquiries. In 2017 he was named Health & Safety Silk of the year at the Chambers & Partners UK Bar awards. His practice encompasses administrative law and civil common law with a particular focus on personal injury.

ONLY HUMAN

SANJAY MUNNOO

The African Nguni Bantu term of 'ubuntu' refers to a quality of essential human virtues compassion and humanity, and it is more important now than ever before.

As the prevalence of COVID-19 has increased in South Africa, it has been challenging for employers and employees alike. The South African Institute of Occupational Safety and Health (SAIOSH), as the only South African Qualifications Authority (SAQA) certified body and the largest entity of OHS professionals in Africa, has been called upon to respond. In the midst of the COVID-19 crisis, however, we must not lose sight of the ongoing toll of workplace injuries.

After reflecting on many of the preventable work-related injuries and illnesses that take place around the world on an annual basis, I note that South Africa is no different, whether it is an injury related to work, home or travel.

When I was younger, I felt invincible and oblivious to any danger. I recall my late dad, Reg, cautioning me about speeding and saying "Thank you" to the driver after arriving safely at our destination. At the time, I didn't fully appreciate his concern, however, his advice fully resonates when reading statistics that over 134,000 people were killed on South African roads from 2008 to 2017.

The unfortunate reality is many young adults are impervious to inherent dangers that exist at the workplace. From experience, I noted an increased proportion of injury and on-duty fatalities among younger workers aged 20 to 30 years, including a growing trend among young women.

The World Health Organization (WHO) defines health as "a state of complete physical, mental and social well-being". Education at our schools has an essential role to play in enabling people to work towards that state, by preparing new generations for their adult lives. I believe health and safety is the responsibility of all citizens. For young children, education should address general concepts of health, safety and well-being, as well as socialization. For teenagers and young adults, the focus should be on promoting the concept of health, safety and well-being at work and in life.

This requires appropriate content to be taught and for educational establishments to model health and safety. Learners should play an active role in the health and safety of their school, take ownership of their environment and understand how it is managed.

In my opinion, there could have been a lower incidence of COVID-19 infections had society been health and safety conscious. I'm a firm believer that demand will continue to grow for capable H&S Professionals post COVID-19 and beyond. The key to preparing our world for whatever the future holds is preparing our youth for a safe and healthy future, and for leaders to remember that workers – whoever they are, wherever they are – are only human.

Sanjay Munnoo has over 15 years' experience in the risk and financial services industries. Sanjay is a Chartered Member of SAIOSH (CMSAIOSH) and was appointed as President of SAIOSH in June 2017. He is currently completing a PhD in Construction Management.

HEAT OR LIGHT?

KEVIN MYERS

My viewpoint on health and safety is framed and informed by my experience as a regulator. What is a regulator? In my opinion, it means being an influencer and a change agent. Its fundamental role is – or *should be* – to influence change in organisations' behaviour and culture in a way that improves safety and health outcomes. But regulators realise that whilst some change because they see the light, others are reluctant to do so unless the feel the heat.

Of course, there are others (such as OSH professionals) with roles to play within organisations that have – or should have – a similar role in driving change. I wonder what OSH professionals might learn from good and effective regulation. My experience is that, more often, any such prior learning is of poor, or simplistic, regulation – which tends to be less effective.

Instead, I'd like to offer a few thoughts on how *good* regulators think:

- The expression "of course I love you" might set off alarm bells of insincerity in a relationship. For a regulator, the equivalent is when an organisation or CEO says "safety is our number one priority". I'm not suggesting people or organizations are setting out to mislead if they use the expression. They may well be deluding themselves. The number one priority for any organisation is survival. Excellent OSH performance will help in this respect (and for many organisations it is a *sine qua non*). But it is not a priority – it is, if anything, a 'value.'
- That value is often expressed by the culture established in an organisation. Most people in an organisation know what its culture is. If good OSH is really valued, it will be part of the culture. Get the culture right and the process and manuals just become reference.
- If good OSH isn't part of the existing culture, how do you go about change that? Some thoughts:
- Find important problems and fix them.
- Think in OSH outcomes, not processes. Worry less about what your systems contain and focus more on what they deliver for safety.
- Always beware optimism bias: Don't ever think "it probably won't happen to me".
- Sell don't tell. Change is more sustainable if people buy into it, rather than just being told to follow the rules.

An organisation's impact on OSH performance doesn't stop at the metaphorical factory gate. Its performance is only as good as that of its supply chain.

Kevin Myers is President of the International Association of Labour Inspection (IALI). Previously, at Great Britain's Health & Safety Executive, he held a wide range of operational, policy and strategic roles including Chief Inspector of Construction, Director of Hazardous Installation, Director General – Regulation, Deputy CEO, and Acting CEO.

SELL DON'T TELL

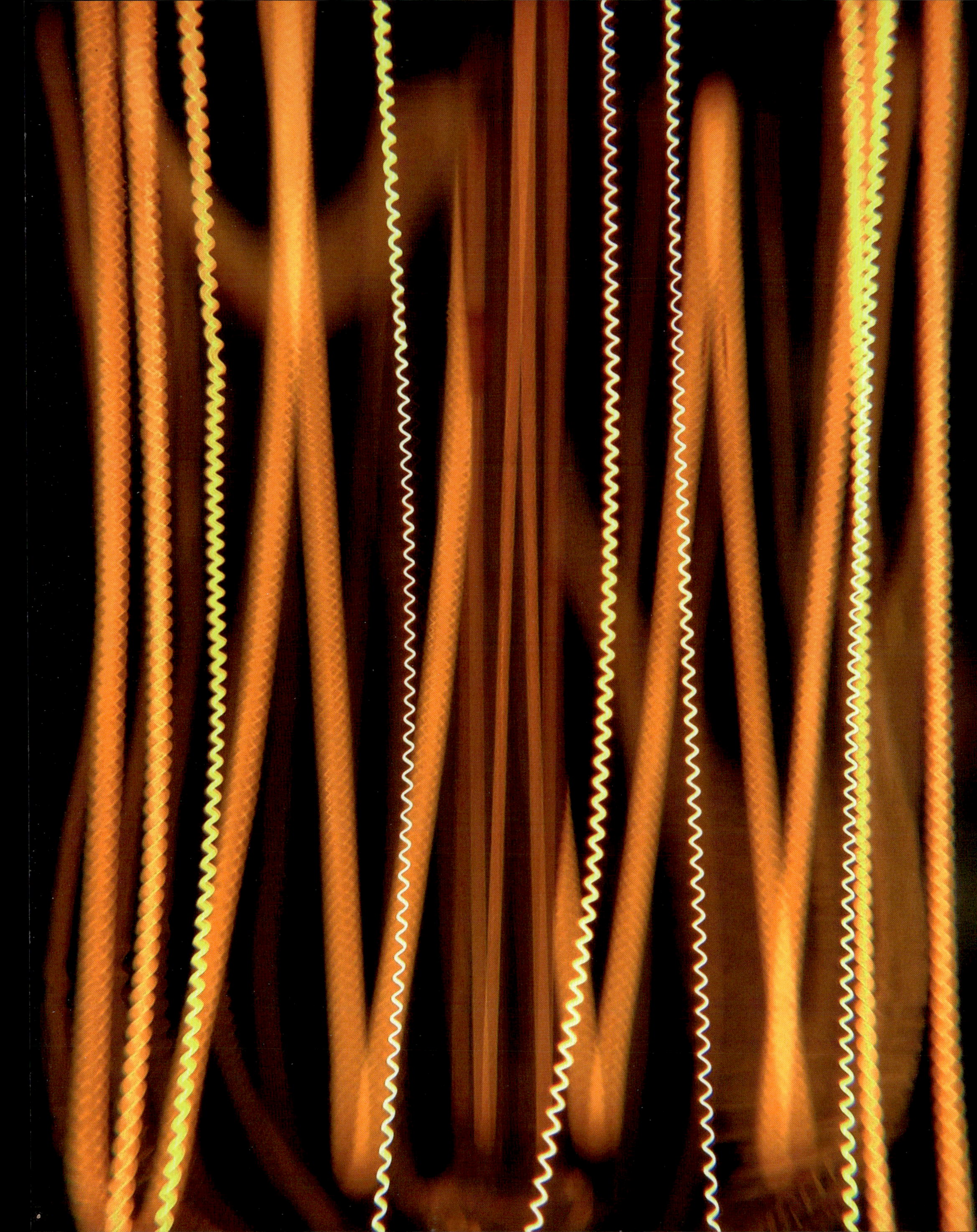

MORE THAN MEETS THE EYE

MARCIN NAZARUK

"Failure to recognize the hazard" is a common root cause in incident investigations.

Leaders often ask: "How could they have not recognized the hazard? If only workers were aware of hazards, we wouldn't have accidents."

But there is more to this discussion than meets the eye:

1. *Identifying sources of potential harm (hazards) is difficult.*

Ability to recognize hazards is linked to broad safety knowledge. For example, in a study on how miners identify hazards, 'the best' front line operators missed 50% of the existing hazards[1]. Safety professionals tend to identify more hazards than frontline workers. Groups which include different skillsets and roles tend to find most hazards.

Practical Takeaway: Don't rely on one person conducting risk assessment but get people together to identify hazards.

Instead of starting with a generic question: "What may harm you on this job?", break the activity into steps, and for each step, discuss how potential energy sources (e.g. electrical, mechanical, pressure, chemical, etc.) could result in harm. This method generated over 30% more identified hazards – according to another study[2].

2. *Judging probability of getting hurt (risk) is a different mental process.*

One study[3] showed that workers make the judgement about the likelihood of getting harmed based on their personal experience. If workers were injured in the past, they feel that that they are more likely to be harmed again. Those who did not have a bad experience with the particular job will say that they are less likely to be harmed.

Practical Takeaway: Introduce exercises simulating different incident scenarios and discuss how they could happen "where we work" to strengthen the feeling "it could happen here".

3. *Whose risk perception?*

Too often, the discussion on risk perception focuses on the frontline employees. What about other people who contribute to the level of risk and how it's managed? This includes: HSE managers, supervisors, operations managers, engineers, planners, procurement and so forth. What role do their perceptions of hazards and risk play in generating and controlling the risk?

Practical Takeaway: Include those individuals in a discussion on how they contribute to creating and managing the risk related to specific activities. Expand your focus far beyond frontline employees.

4. *Simply improving hazard awareness and risk perception is not enough.*

One study[4] found that risk perception did not predict unsafe behaviour, but rather the behaviour was influenced by the working conditions.

Practical Takeaway: Focus your improvement efforts on the situational constraints and performance shaping factors – says the study.

5. *Interpretation of how accidents happen is subjective and constrained by our beliefs.*

We have a natural tendency to attribute 'causes' of incidents to personal characteristics of individuals (see: fundamental attribution error & hindsight bias). "Failure to identify hazard" used as a root-cause hides the details of the situation and prevents effective learning.

Practical Takeaway: Focus on understanding the context and situational constraints to explain behaviour.

1. Bahn, S. (2013) *Workplace hazard identification and management: The case of an underground mining operation*, Safety Science. doi: 10.1016/j.ssci.2013.01.010.
2. Albert, A., Hallowell, M. R. and Kleiner, B. M. (2014) 'Enhancing Construction Hazard Recognition and Communication with Energy-Based Cognitive Mnemonics and Safety Meeting Maturity Model: Multiple Baseline Study', *Journal of Construction Engineering and Management*, 140(2), p. 04013042. doi: 10.1061/(ASCE)CO.1943-7862.0000790
3. Brewer, N. T., Chapman, G. B., Gibbons, F. X., Gerrard, M., McCaul, K. D. and Weinstein, N. D. (2007) 'Meta-analysis of the relationship between risk perception and health behavior: The example of vaccination.', *Health Psychology*, 26(2), pp. 136–145. doi: 10.1037/0278-6133.26.2.136.
4. Rundmo, T. (1997) 'Associations between risk perception and safety', *Safety Science*, 24(3), pp. 197–209.

Dr Marcin Nazaruk is a Global Human Performance and Culture Leader at Baker Hughes. He's held a number of senior leadership roles in the energy industry focused on human performance. Nazaruk has extensive experience in modernizing risk management and safety culture through demonstrating the value of applied psychology.

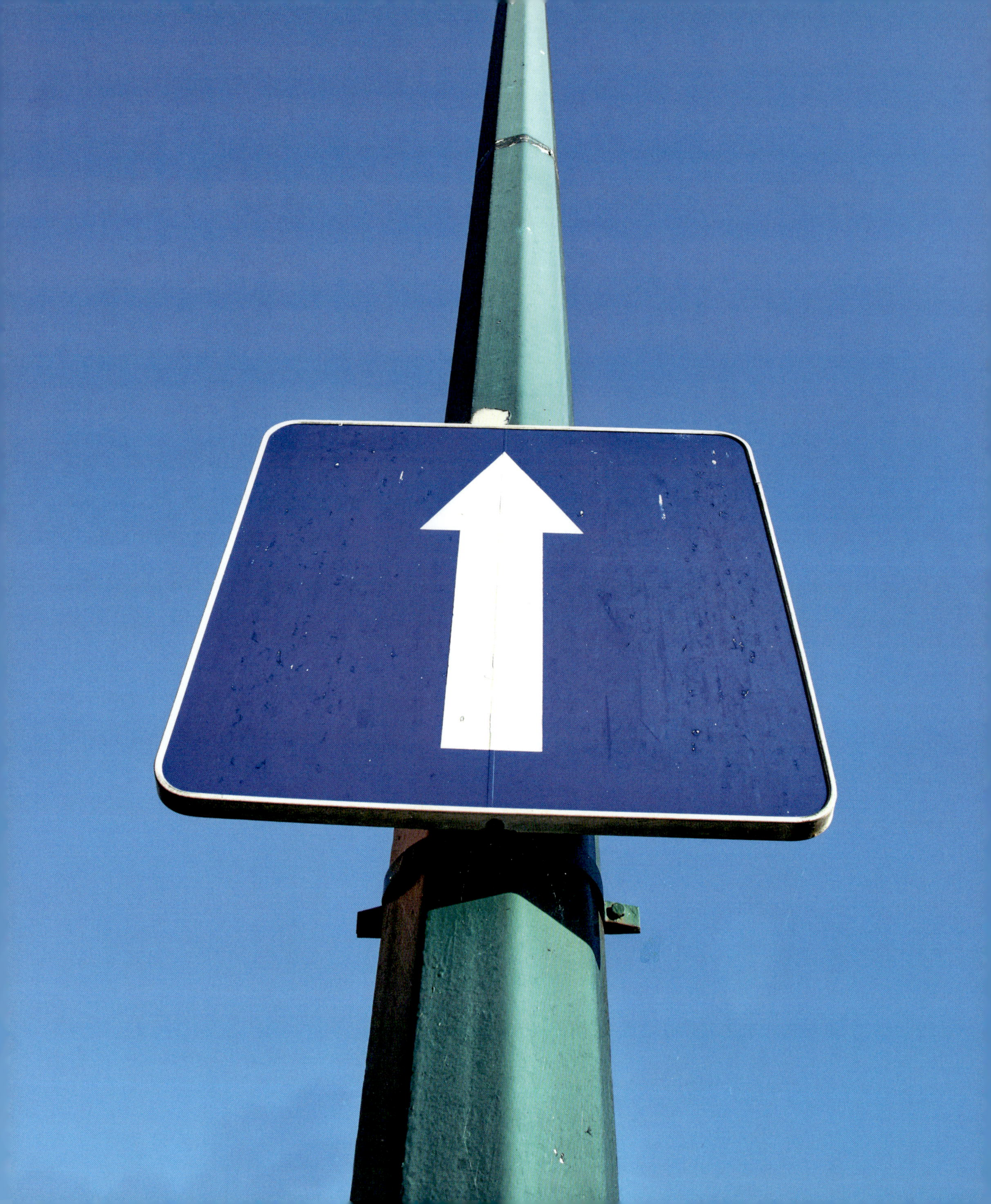

LEADING BY EXAMPLE

MICHAEL O'TOOLE

Safety and health processes within the workplace have taken numerous twists and turns in approach and emphasis, always focused on the intent of saving lives and eliminating mishaps. Too often, an organization changes its approach to managing safety based on the latest fad or apparent silver bullet, which of course doesn't exist.

The latest focus for many organizations is to leverage this thing called 'safety culture' and treat it as if it is the 'holy grail' of safety. "If only we could improve our safety culture, all of our safety issues would evaporate." If only that were true.

A simplistic view of safety culture is the culmination of what management says and how it executes the safety policies and procedures throughout the organization. Yes, the employees certainly have a responsibility within an organization's safety process to actively participate in that process. However, management must visibly demonstrate its consistent support and commitment to the safety and health of its employees, at all levels of management.

Management teams appear to be searching for ways to 'convince' workers to be safe, to not take that shortcut, or perform an 'unsafe act' as they go about their job tasks. There are two powerful tools that management teams can use, but often fail to employ in their efforts to improve safety results. One is known as behavior modeling, or simply known as "leading by example." The other is the use of occasional positive feedback.

Management teams that consciously lead by example demonstrate the importance of safety to the organization, to employees and to themselves. These organizations have experienced lower accident rates, lower turnover, absenteeism and higher employee morale. The interesting thing about behavior modeling, is that it is absolutely free!

Management and supervision are usually pretty good when it comes to recognizing failures. Just ask most any employee, "When you mess up, how long does it take the boss to find you?" That question usually generates a weak smile or smirk and a response of "nanoseconds". However, when asked how long has it been since a boss noticed that you did your job as expected, the answer is a blank stare. A clear way to strengthen expectations is for management and supervision to occasionally take time and "give notice" to employees who perform tasks according to the procedure.

So, what can management do to take advantage of these two simple techniques? First of all, senior management can establish clear expectations for the use of these two tools by their direct reports. They in turn will establish similar requirements on their direct reports. The key here is that management must measure these efforts at all levels, otherwise the suggestion is just that: a suggestion.

When leaders establish these expectations and holds themselves and direct reports accountable, it becomes part of the safety culture. In the end, these efforts send a strong message that has a positive influence on all employees' behaviors. It helps to demonstrate that safety *does* matter.

Dr Michael F. O'Toole is Professor and Program Coordinator of the Aerospace and Occupational Safety Program at Embry-Riddle Aeronautical University. Previously he was OSH Director for two major corporations. O'Toole holds an MA in Industrial Psychology, an MS in Safety Engineering Technology and a Ph.D. in Public Health Policy Administration.

THINGS I WISH I'D KNOWN

DIANE PARKER

Here's a quick and simple way to bring a new colleague up to speed:

Brainstorm the five most useful things that you know now about working safely, but that are not written down or included in the training you get.

Things that everyone gets to understand once they have been with the company a few months. Things that keep everything running smoothly and safely. Come on, every company has a set of unwritten rules and assumptions … this is "the way we do things around here" that gets talked about in seminars on organizational safety culture.

It could be about where the well-stocked PPE locker is, or who the people are who will try to persuade you to bend some of the rules, or how to find an electronic copy of the procedures relevant to your task.

It could be anything, but it definitely will be the things you really need to know to work safely, the things you wish you had known from the day you started.

Once you have agreed on them, get these five key points written up on cards and laminated. Keep it simple, keep it real.

It should be the size of a playing card so that it will fit into someone's pocket.

Make sure that you give a card to every person who starts work in your team. Go through it with them. This is especially important if they are only going to be with you for a short time: a shift, a week, the time it takes to maintain a piece of specialist kit.

In doing this you will show that:

- You have thought about the job in hand.
- You care about your colleagues' safety.
- You want to welcome them as a new member of the team.

Professor Dianne Parker is an applied social psychologist with over 30 years' experience researching the psychological aspects of safe behaviour in a range of high-risk industries. She was one of the original team that worked on developing the Shell Hearts and Minds toolkit.

Brainstorm the five most useful things that you know now about working safely, but that are not written down or included in the training you get.

RISK MANAGEMENT & COMPETENCE

PHILLIP PEARSON

Today, more than ever, from regulators to customers to senior management, organizations are under pressure to be able to clearly articulate how they identify and manage risks.

Fundamentally, everyone is a 'manager of risk' and has an important role to play. Whatever your role, you make decisions that involve an element of risk.

We know already from the UK government's review of the Grenfell Tower fire that more holistic approaches to risk management would have paid dividends in preventing the initial fire and reducing the effects of the disaster to follow. Instead of looking at building materials, construction techniques, fire protection, life preservation, brigade access, evacuation arrangements and building maintenance and alterations separately, it is now recognized that a more holistic 'joined up' approach would have helped greatly. Individual risks could still have been assessed and mitigated, but, critically, we also need to consider the way they interact.

Effective risk management requires a combination of specialized risk expertise, awareness and competence that underpins organizational policies, processes, cultures and leadership at all levels. Therefore, no matter our chosen area of work, whether it be health and safety, finance, compliance, HR or facilities, we all have a part to play in identifying and managing risks to keep people safe, manage costs, meet stakeholder expectations and enhance profits and reputation.

Business leaders who implement and use a competence framework can improve how their organization manages risk by creating a common approach to identifying and managing risks and opportunities at all levels.

Engaged employees who understand their role in fulfilling strategic objectives are extremely valuable and provide insight that may otherwise be overlooked. It can also be a useful tool for teams and individuals to assess and develop their capabilities, such as to:

- Provide a consistent approach to staff development across an organization
- Enhance employee and organizational effectiveness in identifying, communicating and managing risk
- Assist with workforce planning and succession planning with the identification of employee learning needs, from starters to high potential talent
- Embed risk management competence into job roles, providing a consistent approach to recruitment and performance management
- Provide a pragmatic vehicle for open dialogue between management and employees at all levels for effectively managing risk
- Target training resources more effectively and encourage individuals to take responsibility for their own development
- Integrate risk management into existing organizational policies, procedures, practices, cultures and systems

Indeed, a competence framework can be highly effective in helping organizations be more efficient and cohesive by breaking down traditional professional boundaries and developing more flexibility between teams and functions. Business leaders should be prepared to be on a continual learning curve – utilizing the skill sets within their teams to best effect. Using a risk management and leadership competence framework to begin the conversations that matter will ensure you are engaging with your people and helping your business to thrive. You can review or download IIRSM's risk management and leadership competence framework free of charge from their website.

Phillip Pearson has over 25 years' experience within professional bodies from a range of disciplines, including engineering, taxation and surveying. His primary focus is on member education and services. Pearson was appointed CEO of the International Institute of Risk & Safety Management in 2013 and presided over significant strategic change.

CROWDED OUT

SALLY PERCY

Crowds are bad for our health, and they always have been. Throughout time, city-dwelling populations have tended to bear the brunt of epidemics. The World Health Organization says that outbreaks of disease "are more frequent and more severe when the population density is high".

The world has changed dramatically since the 14th century, when the Black Death – the most fatal pandemic in human history – wiped out up to 60% of Europe's inhabitants. Nevertheless, the COVID-19 pandemic of 2020 highlights that we are not so far removed from our ancestors as we like to think.

Certainly, we have technology these days, technology that enables us to do many amazing things, not least work from home. Despite this, however, our survival strategy for defeating COVID-19 has been largely centred on us hiding away in our houses. It's an effective strategy, but a primitive one.

Fundamentally, COVID-19 has reminded us of the risks that overcrowding poses to public health. Yet people all around the world continue to live and work in overcrowded conditions. Significantly, mass transport systems tend to be particularly overcrowded.

On the London Tube, for example, people can be crushed together in hot, airless carriages for up to an hour – heads under armpits, breathing in each other's recycled air. How can this be safe? If it were animals travelling in those conditions, the practice would have been outlawed years ago.

So, here's my suggestion for making the world one percent safer: We need to turn rush hour into a thing of the past. That means investing in our mass transport systems to increase their capacity and permanently changing the working practices of our organizations, so that 9-5 is no longer the default for everyone. The best way to reduce overcrowding on public transport is to encourage people to travel at different times, but they need their employers to make this possible.

Regardless of whether people are travelling on a train, a tram or a bus, let's commit to giving them all a seat. That might not be the equivalent of providing them with a metre or two of dedicated personal space, but it will reduce their contact with other travellers so they are less likely to pick up bugs. Having a seat will also minimize the risk of commuters fainting due to the heat and reduce the general stress they suffer as a result of the journey.

Overcrowding happens in many different situations, with the problem inevitably more pronounced in highly populated countries. Transport is not the only means of addressing the challenge – but it is a very good place to start.

Sally Percy is a business and finance journalist, editor of *Edge*, the official journal of the Institute of Leadership & Management, and a contributor to *Forbes*.

THERE'S NOTHING MORE IMPORTANT THAN YOUR LIFE AND YOUR SAFETY

DO YOU REALLY 'LIKE' SAFETY?

GERRIT POGGENPOHL & CHRISTIAN BOEHMER

You know this from your everyday life as a manager: with the abundance of demands, it is not easy to give proper attention to everything, everyday. This quickly becomes apparent in safety work. Employees can sense whether a statement of intent is only said, or if it's actually lived. The slogan *'Safety First'* is a polarizing example of this. Is it <u>really</u> safety that always comes first? *Always?*

This graphic can help us visualize the controversy between the message as intended, as demonstrated and the message as received. The manager turns his back on the employees who are at risk of an acute explosion while in parallel social media attention points are distributed by 'likes'. Above all the slogan *'Safety First...there's nothing more important that your life and your safety.'*

Emotional messaging works. We like to feel liked. We know that emotionalizing aspects of safety and then communicating them can be effective. Though in addition, the sincere attention of managers is required to resonate and reinforce what's most important. Of course, emotional communication doesn't negate the need for preventative measures, which are crucial. For example, Near Miss reporting helps to learn from events that have not led (yet) to an accident.

Dr Gerrit Poggenpohl is Head of Global EHSQ & Global Product Regulatory for BASF Construction Chemicals. He holds a PhD in Process & System Technology.

Christian Böhmer has his roots in the graffiti movement of the late '90s -still a leading influence on his style. His current focus is portrait painting.

SEEK TO UNDERSTAND

JAMES POMEROY

Every day I travel to work in London, I pass a reminder of our mission in OHS and what we are trying to protect. It's a 300 kg bronze statue of a building worker in a pose reminiscent of Michelangelo's David, but dressed in work clothes, wearing a hard hat and holding a spirit level. This statue was erected to commemorate the building workers who lost their lives at work and to celebrate all those involved in construction sector. It has become even more pertinent to me during the pandemic, when we have rediscovered importance of frontline workers who often put themselves in harm's way every day, and whose work is often taken for granted.

Passing the statue, I am reminded of the need to listen and learn from the frontline workers doing the work. This sounds obvious and doubtless a little trite, but much of the everyday activities that a modern safety professional undertakes actually distances us from engaging with the very people that we trying to protect. Making time to listen from the *real* safety experts, the workers, has never been more important.

There are many things that we need to do to address the horrific figure of 2.78 million work-related deaths and 374 million non-fatal work-related injuries that occur each year. One of the most important is a better appreciation of the difficulty workers face when trying to apply the safety solutions, theories and processes we create. And when things unfortunately do go wrong, we must apply the same degree of curiosity when seeking to understand the context, challenges and demands that the individuals where facing. As Stephen Covey observed in his book *The 7 Habits of Highly Effective People*, we must seek to understand before seeking to be understood.

As I stare up at the statue, I wonder about what the worker could tell us about keeping safe, and this reminds me of the refreshing new views of safety. Recognizing that safe outcomes are often dependent upon the individual adjustments and modifications that workers make provides another reason to engage and learn from the workforce. How much focus, I wonder, do we each give to genuinely seek to learn from their ideas, experience and innovations.

We are at a pivotal point in the development of occupational safety, with many new theories and technologies being proposed. The world of work is fundamentally changing, and there is an increasing recognition that performance is plateauing. As we seek to address these challenges, we must not overlook the importance of listening and learning from those we seek to protect. We owe that to the 2.78 million workers who don't return home each year.

James Pomeroy is the Group HSES Director for Lloyd's Register. Pomeroy has been involved in leading and transforming HSES programs in a variety of sectors and countries for over 30 years. He is a Fellow of IOSH and IEMA, and holds degrees in engineering, business management, environmental law and OHS.

FINDING COMMON GROUND

DREW RAE

There are many opinions and debates about safety, but hopefully, there are a few things that we can all agree on.

We can agree that the safety of any work depends on the physical environment in which that work is performed. Working at heights or by the side of a road is less safe than performing the same work in a dedicated, protected space. Flying a Comet airliner in the 1950s was less safe than flying a Boeing 737 in the 1980s.

We can agree that the safety of any work depends on having the right tools for the job. Software safety experts argue about programming languages and formal methods for the same reason that construction companies restrict the use of box cutters and angle grinders: because they believe that the choice of tools matters.

We can agree that the safety of any work depends on having the right people around. Working physically alone or in a hostile social environment is more dangerous than performing the same work surrounded by people willing to monitor, assist, and sometimes, to constructively challenge or correct the work.

We can agree that the safety of work depends on how the work is done. There is no 'one best way' to perform any given task, but there are certainly safe ways and unsafe ways to do a job.

Investing in any of these four things – the physical environment, the tools, the people, or the practices – will almost always make work safer. So why do we spend so much time, effort, and attention on things which, at best, only indirectly influence the safety we all agree on?

Our systems, procedures, presentations, reports and risk assessments are not safety. They do not cause safety. They – maybe – *cause the things that cause the things that provide safety*.

My suggestion for one, high impact thing that you can do to make the world safer is to ask your workers which safety practice they find most annoying. Then, try to find evidence that it genuinely makes work safer. If you can't find the evidence, create a fair test to produce the evidence yourself. If you still can't demonstrate that the practice improves safety, get rid of it.

The best thing about this process is that that there is already good evidence that decluttering safety work in this way improves the safety of work. By asking workers questions, we might also find common ground and common language between managers, OSH practitioners and the frontline.

Dr Drew Rae is Senior Lecturer at Griffith University, Australia. Rae's research uses ethnography and field experiments to investigate myths, rituals and bad habits that surround safety work, and how these influence frontline operations. He co-hosts the "Safety of Work" podcast and is Associate Editor for the journal *Safety Science*.

BUILDING A CULTURE OF HEALTH

ANTHONY RENSHAW

We are uniquely positioned right now to maximize on the biggest opportunity organizations have had in a generation to affect meaningful change on the health of their employees. While COVID-19 may lead some to sedentary lifestyles, stress and anxiety, alcohol and smoking, in others, we may see evidence of increased social cohesion, better sleep and adoption of more healthy habits. It remains to be seen what the long-term effects will be, but they will be significant.

In 2016, we saw the first drop in life expectancy in 20 years. Risk factors and healthcare access all contribute. Obesity and undernutrition are, perversely, on the rise in many countries; this will continue after COVID-19, with its resultant socio-economic changes. Seven in ten top causes of death in most countries are caused by a chronic disease – and this will affect developing countries increasingly. The impact of closures of health facilities on workers' long-term conditions, as well as vaccination rates, is likely to be high.

So what can business do about all of this?
The answer is: a lot. Solving these issues will need to engage businesses of all types. Why? Because we know that businesses also get hurt as a result of poorer health. If you have a less healthy workforce, the cost of providing health insurance may be much higher. You probably experience more absenteeism among your workers, or presenteeism, where your workers show up while they're sick and can infect other employees. All of this will continue unabated post COVID-19, unless there is meaningful change.

On the other hand, we have growing evidence from around the world that business leaders who care about health can turn that into a positive business strategy. For example, research demonstrates that high sustainability companies significantly outperform their business competitors over the long-term. These companies use performance reporting to their boards as an essential element of corporate governance. We also know that the share prices of companies that prioritized health outperform the S&P average on all tests. How do they do this? My view is that their strong health leadership is making health a priority – just as safety has entered the mainstream.

Businesses have to support health service provisions in locations with limited services; some have already provided exceptional support to employees affected by outbreaks; some transform the incidence of malaria and TB in their communities; some have developed world-leading global wellness programs.

These are all an example of a 'culture of health': one where individuals and businesses can make healthy life choices within an environment that promotes wellbeing options, where the healthy choice becomes the valued and easy choice.

What COVID-19 has shown us is that all businesses are in the business of health. I sincerely hope that the future will show just how positive an impact business can have on health through a focus on creating change – to the health of workers, their customers, their communities and the environment.

Dr Anthony Renshaw BSc (Hons), MB ChB, MD (Res), MRCS (Ed), MBA, FRSPH, FFFMLM is Medical Director, Health Consulting, Europe for International SOS. He provides medical advice to global organizations and has consulted across Africa, Asia and Europe. Prior to joining International SOS, he was a surgeon in the UK and South Africa.

EMBRACE DYSFUNCTIONAL TEAMS

ROB RICHARDSON

We are all familiar with the faded poster of safety rules: by the water cooler, on the factory floor, in the mess room. Rules have tarnished safety at work for years. Though some people like their 'black and white' clarity, leaders can also hide behind rules.

Rules can become a barrier to engagement. Instead of adhering to tired standards, in modern organizations, we should *want* people to have opinions, thoughts and ideas to make their own judgements.

We fill our workplaces with signs… 'do this', 'don't do that', 'watch out for this', 'this could happen if', and on. These are well-intentioned, but in reality, we might as well erect a sign that says" 'We don't trust you!' I have seen far greater success where people are given the tools, the training and the confidence to work safely without bombarding them with rules.

This focus on rules, among other factors, has contributed to dysfunctional teams. While dysfunctional teams are often seen as problems to solve, they actually offer an important opportunity. I see dysfunctional teams as a potential breeding ground for debate, opinion and ideas. Of course, it would be great to say 'we have developed a culture that achieves this naturally', but we live in the real world, not the perfect world, and our dysfunctional teams are a by-product of that.

So, rather than supressing people and try to get them to conform, let's ask them what they think would make the organization better and safer. We might ask: *What rules don't make sense to you? What would you change about your team if you were in charge? What would enable you to do your job better? What makes you feel unsafe?*

Let's accept that everyone is different, sometimes pulling in different directions. Instead of trying to make them uniform, let's listen to what they have to say.

It might save someone's job. It might even save someone's life.

Rob Richardson worked 31 years for Heineken in the UK where he was Production Director, in New Zealand as Supply Chain Director and his final role was in the Netherlands as Area Supply Chain Director for Europe.

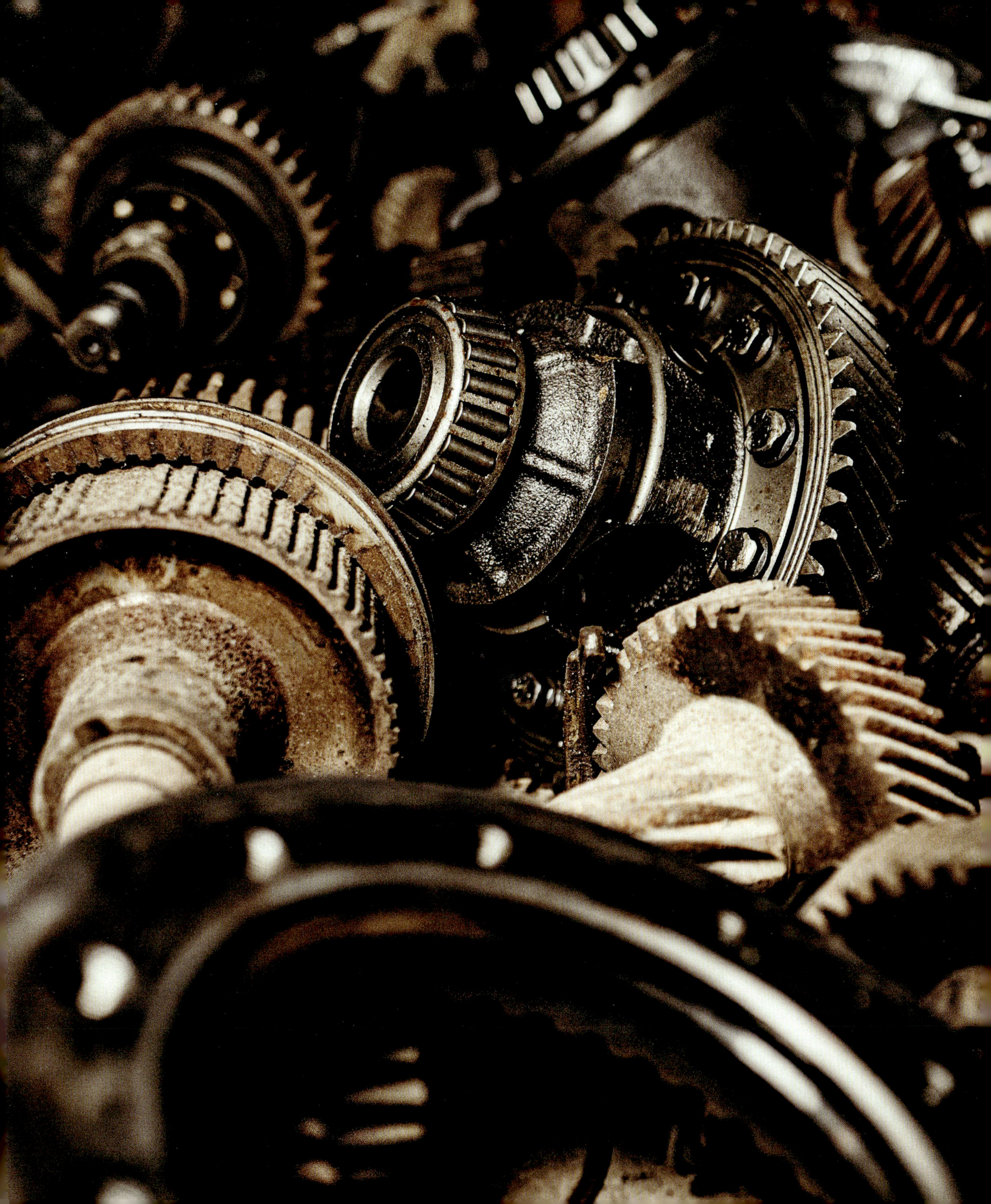

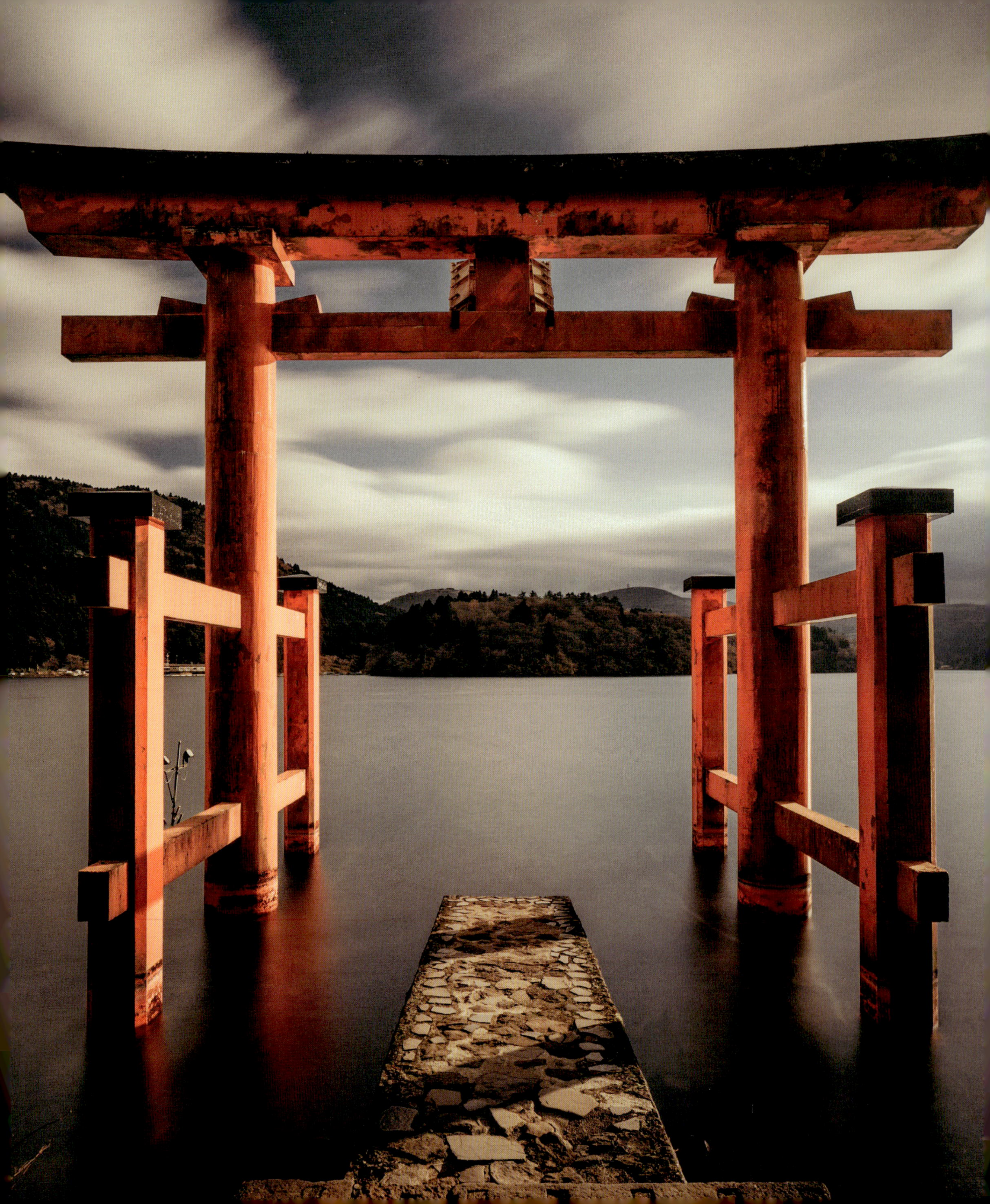

HAIKU FOR QUALITY CONTROL

NORMAN RITCHIE

Recommendations?
What will save lives is action
Not aspiration.

The haiku form of poetry originated in Japan around the sixteenth century. According to tradition, a poem must meet well defined structural requirements to be classed as haiku. In other words, it must pass a quality control test.

As with poetry, so with incident investigation…

When something goes wrong during an activity, let's call it an unplanned event, the entity concerned generally doesn't want it to happen again. This is especially true if significant consequences occur as a result of the unplanned event. Time and treasure are expended on an investigation, which can be boiled down to three components: information, analysis and action.

If the goal of an investigation is to significantly reduce the probability of a repeat of the unplanned event, why do so many incidents reoccur? While the answer to that question is complex, one stumbling block stands out: the critical transition from recommendation to reality. Most safety and risk management professionals are familiar with cause and effect methodology, but few are trained in quality control of the output of their analysis.

If the way an activity is conducted in real life, tomorrow, has not changed from the way it was done today, nothing has been done to prevent the unplanned event from occurring again. To prevent reoccurrences, the output of cause and effect analysis must lead to a change that is both relevant to, and effective against, the unplanned event. Too often, recommendations generated during analysis remain unimplemented, as excellent and wise as they may appear to those involved. No risk reduction has been achieved until recommendations have been approved by an adequate and appropriate authority, and then – most importantly – implemented.

Front-line safety and risk practitioners are not alone in their struggle against this challenge; the same barrier is faced by non-regulatory government agencies charged with investigating the worst of disasters. Impressive investigation reports that talk about how things 'should be' do not save lives or prevent injuries.

Save lives by learning from unplanned events. Validate incident investigations by applying quality control to their output; verify that aspirations have transitioned to a changed reality.

Life-saving actions:
Relevant, effective and
change how work is done.

Norman Ritchie grew up in Scotland and now lives in Houston, Texas, where he applies his many years of project and risk management experience as a Director and Principal Consultant at vPSI Group, LLC.

NO HEALTH WITHOUT MENTAL HEALTH

IVAN ROBERTSON

"There can be no health without mental health."
— Ban Ki Moon, United Nations Secretary General, October 2010

When it comes to mental health, work can be a force for good – but this only applies to 'good work'. Good work is work that is demanding, meaningful and provides workers with the necessary resources to do their job. Challenging work demands are actually healthy, otherwise, how can you feel any sense of achievement? So, a degree of pressure is fine but there is a difference between 'challenge' and 'hindrance' pressures. Pressures that provide a fair challenge can motivate people and help to make work meaningful and satisfying. Pressures that hider performance and make it harder for people to achieve their goals have the opposite effect and create undesirable stress.

People can manage challenging job demands much better when they have a good degree of control over how they do their work, and they have the right resources and support to do the job. If demands are high but control is absent or resources are not there, then it becomes much more difficult to feel on top of things – and that's when mental health and work performance can start to suffer.

A more detailed view of the workplace pressures reveals six sources of pressure:

- Resources and communication (Pressure from lack of resources or information)
- Control and autonomy (Limitations on how the job is done or freedom to make decisions)
- Balanced workload (Peaks and troughs in workload, difficult deadlines, unsocial hours, work-life balance challenges)
- Job security and change (Pressure from change and uncertainty about the future)
- Work relationships (High pressure relationships with colleagues, customers, bosses)
- Job conditions (Pressure from working conditions or pay and benefits)

The consequences for workers and employers when workplace pressures are managed effectively are very positive and lead to healthier levels of psychological wellbeing. When workers have high levels of positive psychological wellbeing, they are more productive and their physical health is protected; the incidence of cardiovascular disease is much higher for workers who experience higher levels of job strain. Even worse, in extreme circumstances, workplace pressures may contribute to suicides.

The people who can make the biggest difference here are the leaders of organizations and their management. There is evidence that leadership and management have an impact on employees' wellbeing, even when controlling for other factors, such as support from home, stressful life events and support from others at work.

What is the most important thing for managers to do for the mental health of their workgroups? Understand how the six sources of pressure outlined above are affecting their teams and do everything possible to get them into proper balance. In other words, challenging demands should be coupled with control, support and resources, security, good relationships with others and healthy working conditions.

Dr Ivan Robertson is co-founder of Robertson Cooper Ltd., a company specializing in mental health and wellbeing, and Emeritus Professor of Work & Organizational Psychology at University of Manchester. He's in the top 2% of most influential psychologists in his field. Robertson's latest book (co-authored) is *Well-being: Productivity & Happiness at Work*.

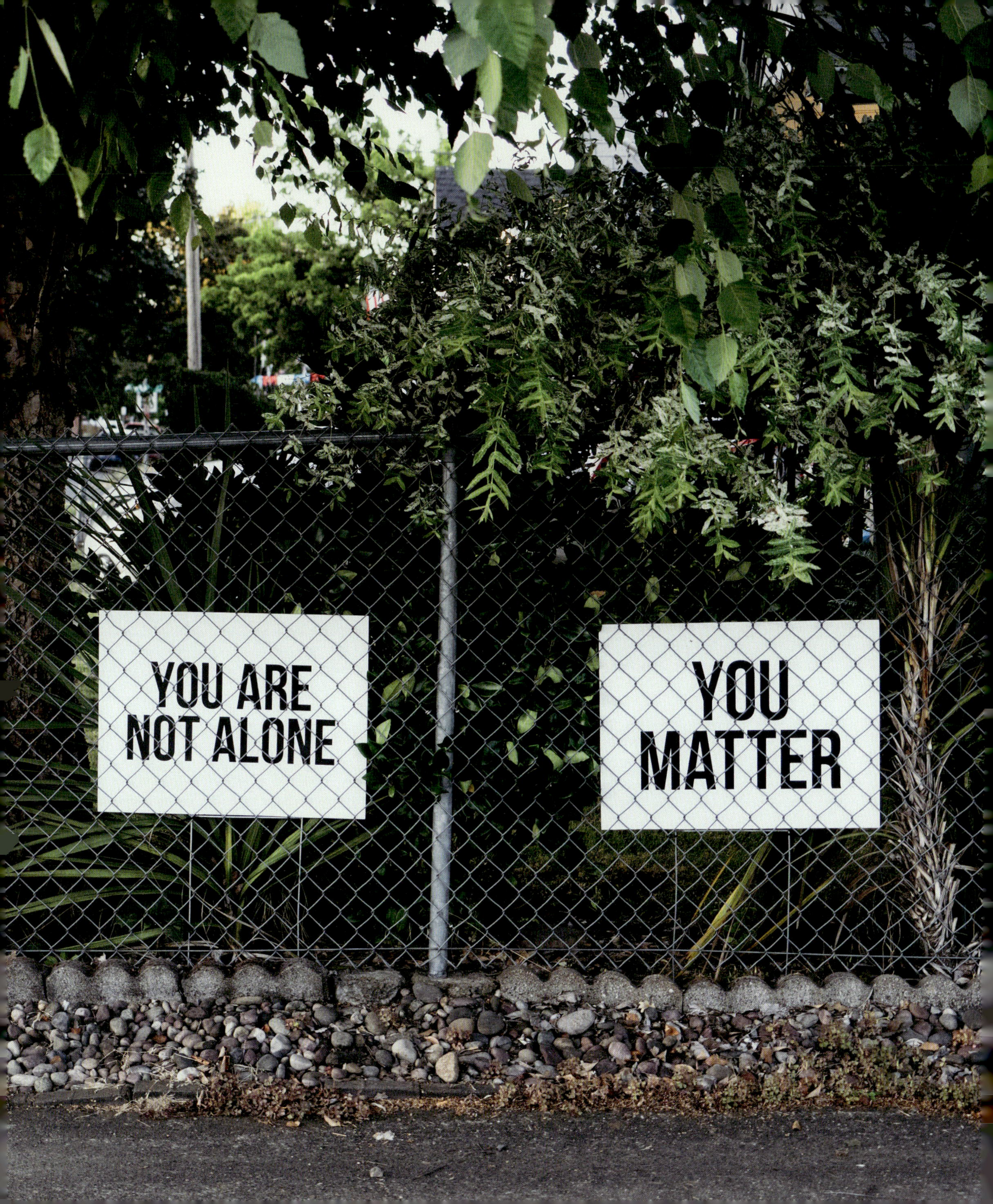
YOU ARE
NOT ALONE
YOU
MATTER

GO/NO GO

FREDRIK ROSENGREN AND ERIK MATTON

Here are two simple questions for you:

When you're out driving, do you always stop at a red light?

When you see a 110 km/h sign, do you always drive at the speed limit?

If you are like most people, you answered *YES* to the red light, and *NO* to the speed limit sign. Take a second and think about why.

If you figured it out, then congratulations! You just cracked the code on how to create desired results, like a strong safety culture, in your organization! You have access to the most powerful tool you need in order to propel behavioural change.

Some of you might want an explanation though.

All results in an organization are created by behaviours. When you see a lot of desired behaviours, you get the desired results. With unwanted behaviours, you get unwanted results.

So, if you are aiming for a strong safety culture, you need a lot of safe behaviours in the organization. How is that achieved?

Like some of you have already figured out, motivation to perform a certain behaviour is powered by the consequences of the behaviour. The red light and the sign didn't make you choose your behaviour; the consequences did. Like risking a collision if you don't stop, or the fact that it will take you longer to reach your destination if you drive at 110 km/h.

Fredrik Rosengren is an engineer and **Erik Matton** is a psychologist. They are co-founders of Behaviour Design Group, which offers strategic expertise and support in organizational change through an evidence-based model and via research on leadership linked to safety culture at the University of Gothenburg.

If a behaviour is followed by a positive consequence, it will probably be performed again. It's called positive reinforcement. The only thing that's certain when you are driving at 110 km/h is that it will take you longer to reach your destination. If you drive at 120 km/h you'll get there faster. And negative consequences, such as a speeding ticket, are rare.

If a behaviour is followed by a negative consequence, it will most likely not be performed again. This is called punishment.

So, if you make sure to reinforce safe behaviours in your organization, safe behaviours will increase. When safe behaviours increase, your safety culture grows stronger and your incident rate will decrease. It's simple math, really. Try it yourself: it's science.

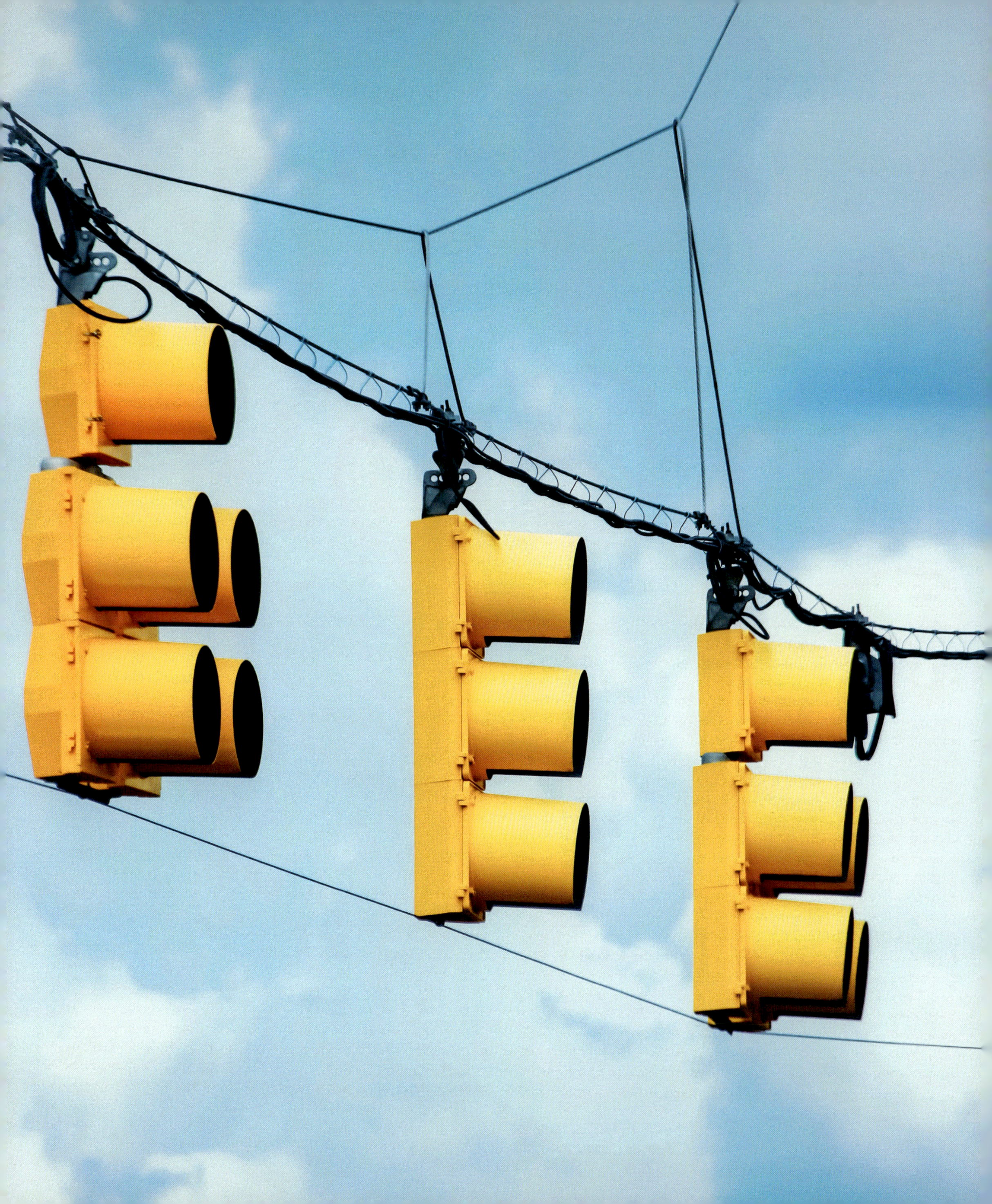

SAFETY STAYS

ACTION, NOT WITNESS

DEBORAH ROWLAND

Deborah Rowland is a leading global thinker, speaker, writer and coach in the field of leading change. Most recently, she co-authored *Still Moving: How to Lead Mindful Change*. She has personally led change at Shell, Gucci Group, BBC Worldwide and PepsiCo, and pioneered original research in the field.

I used to work at Shell. An organization that took – and takes – safety very, very seriously. To this day, I look to take the handrail (hopefully, available) when walking up and down a staircase, because of the emphasis on safety from those years.

But it was at Nice airport 10 years ago that I had my greatest safety lesson. I was at the baggage reclaim, awaiting my bags. Looking forward to a few weeks in the sun. I noticed a very young French child, noticeably bored, pulling away from where his parents were standing. He went to the luggage belt, that was moving, and started to poke his fingers into it – giggling, yet looking back over his shoulder at his parents as he knew he should not be doing it.

I then noticed a tall dark-haired man walk over to him and lean over, gently taking the *le petit garçon's* arm and guiding him away, speaking to him in French. It was all done very quietly yet firmly; the boy did not appear chided or angry.

It then struck me – the tall man who had safeguarded this boy's safety (and indeed others, as all the other children around the belt were looking to join in the fun) was a retired senior Shell executive who I had worked with extensively 15 years prior on a major restructuring of the European business.

I felt humbled. Why had I not gone up to the child to warn him of his dangerous behaviour? *Safety requires action, not just witness.* Whereas Paul, despite being long since retired, still had it in his bones to be alert to and be a guide to safety – wherever he was, and whatever he was doing. He carried safety with him.

It's not something you turn on and off.

Safety stays.

BUILDING COMMUNITY WITHIN YOUR CULTURE

DAVID SARKUS

My thoughts regarding community go back 25 years and evolve from my study of servant leadership.

During that time, a leader at TRW, Inc. – an automotive and aerospace corporation – suggested that he wanted to build a "sense of community" into our safety efforts. From that point, I couldn't put the thought of community to rest. My research and application of community started at TRW, through servant leadership, and what anthropologists identified as "important dimensions of community." In short, community is about people coming together, with shared beliefs and values, and in this context, also with the desire to keep each other safe.

What factors are important for a healthy and safe community?

Mission-Oriented
The great companies I've worked with ensure there is a consistent, mission-like focus on safety. This focus originates from a regular and pervasive form of communications and actions that show that every job must be completed as safely as possible — from start to finish. Safety's a fundamental part of production!

Diverse Views
Communities are built from diverse thoughts, backgrounds and disciplines. This is a community in which many individuals are willing to share different ideas and experiences. Importantly, leaders seek out and nurture different views as the foundation of effective problem solving and innovation.

Openness
Organizational leaders work toward creating openness and honesty regarding hazards and risks by communicating with each other about safety-related activities, giving positive and corrective feedback amongst workers *and also* among leaders. Openness leads to reciprocal openness, transparency and honesty, as well as psychological safety, whereby people feel free to speak their mind without fear of rejection or retaliation. Ongoing openness also leads to increased levels of trust that move groups together in concert, rather than in conflict.

Fairness and Cooperation
Conflict often occurs when there's a perceived lack of fairness, possibly related to discipline, workload, inadequate tools or poor working conditions. Being fair to workers and cooperating for the sake of safety is crucial to sustaining a sense of community. Conflict is part of every organization, but when conflict is healthy, everyone's attention can be aligned back to an enhanced mission for safety improvement. When conflict does persist, there's a "collective conscience" to call upon, and be guided by the organization's values.

Consensus Building
People come together to openly express their views without being judged by others, and agreement should be focused on how a job will be accomplished with risks and concerns identified, as well as acceptable mitigation strategies. Although there may be differing viewpoints, people are encouraged to listen to one another, and move forward once consensus is built.

You can quickly put these five dimensions of community to work within your tailgate meetings, safety chase-downs, timeouts, hazard assessments, senior level meetings or at any time you bring individuals together to improve safety performance. Servant leadership is known to have a profound impact on an organization's safety performance, and community is a vital component of servant leadership.

David Sarkus, MS, CSP is a consultant and leadership coach with over 30 years' experience in safety performance improvement. He holds master's degrees in safety management and organizational psychology. Sarkus has authored five books and been recognized by *Industrial Safety & Hygiene News* as a Top 50 safety professional.

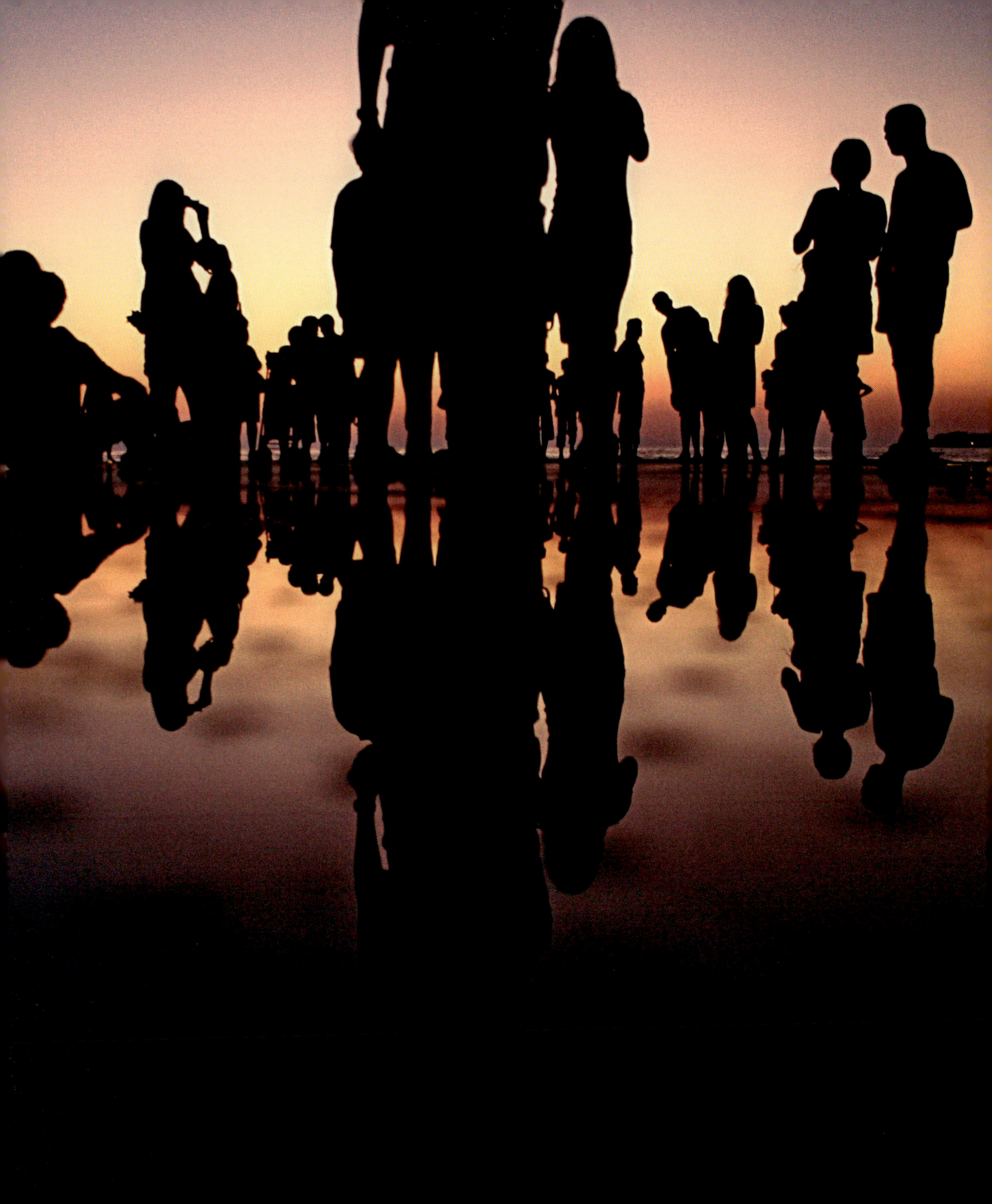

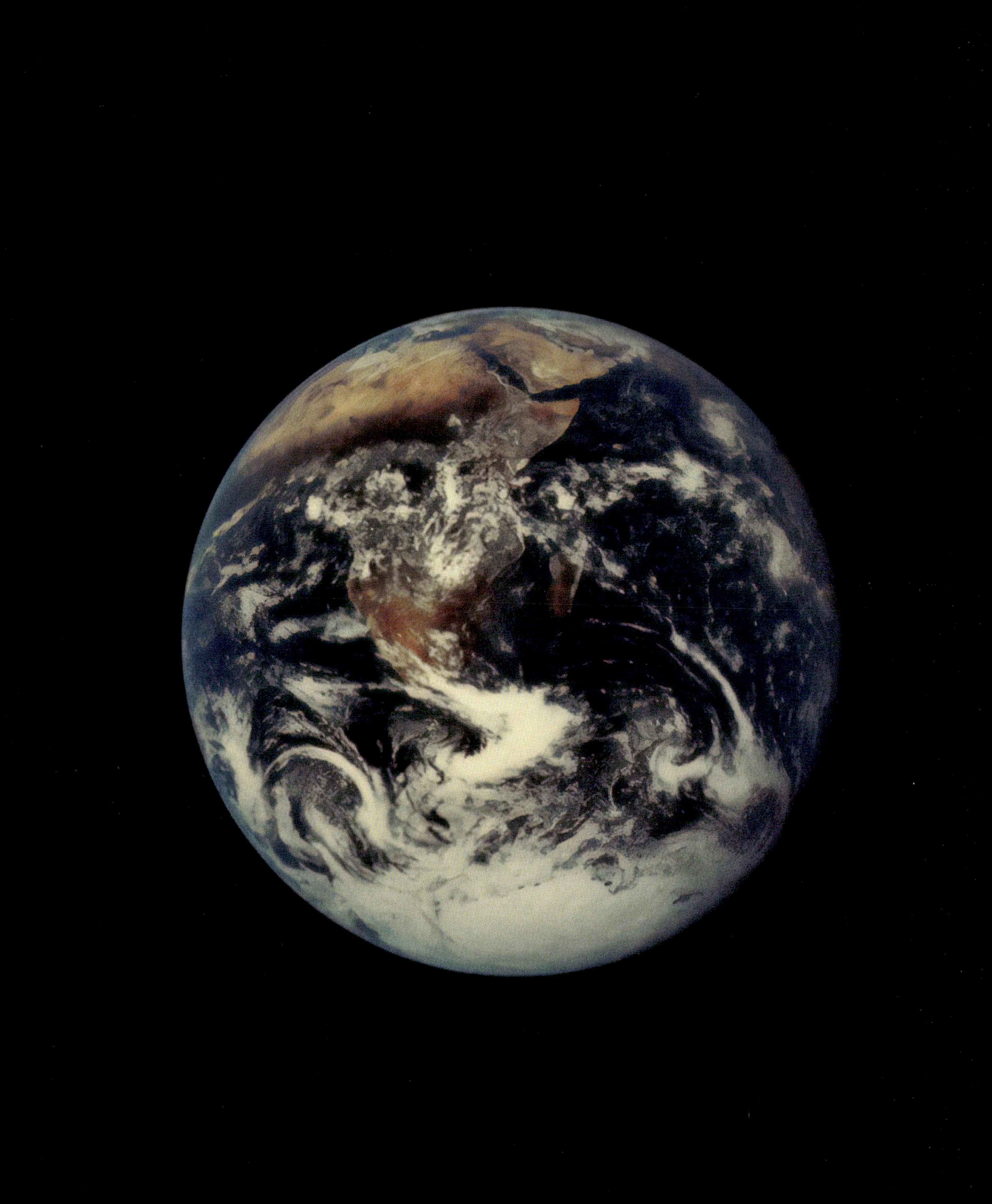

THE HARDEST PART

EDGAR SCHEIN

Dear Group Leaders

As your global CEO, I have called together all of you CEOs and COOs of the major units of our worldwide conglomerate for our usual strategy review. I am also interested in what each of you are doing about *safety* in your organizations. I already have all your accident statistics, but I need to find out several additional things. Please tell me:

1. In the last six months, how many employees in your organization died in relation to their work?
2. What were the circumstances that caused their death? I don't want the 'root cause,' I want to know the complex elements that conspired to cause the death.
3. Did you personally visit the family of the deceased to explain what happened and to find out what the family now needs as a result of the loss of this member?
4. Did you follow up with your own administrators to ensure that the family gets what you discussed with the spouse?
5. Did you bring your direct reports to a meeting to discuss the causes and consequences of each death, and did you as a group discuss how the work could be organized better to avoid such deaths?

Hopefully you already have the answers with you for this important issue, but if you don't, we will go over this in detail on day three, which gives you time to get information from your home office.

I know that you all accept our espoused and publicly posted value that "Safety is Number One," but I want us all to become more aware of the human consequences of death, quite apart from their economic consequences. I have seen too many companies become so obsessed with the accident statistics that they sometimes don't even count the occasional death as part of their safety program. I have heard some of you sometimes say, quite indifferently, that "a few deaths in our high hazard industry is just the cost of doing business."

When I was in your role, I found that there was nothing more sobering than to go to the home of a dead employee and explain to the family what happened. I also found that there was nothing more irresponsible than to avoid this duty and either say nothing or send someone else to break the news to the family. Organizations are *human systems* and must learn to behave that way.

I will want serious answers to the above five questions and will follow up with each of you in the next week to go over them with you, and I especially will want to know what you are doing around Question 5.

Any Questions?

Dr Edgar H. Schein is Professor Emeritus at M.I.T. Sloan School of Management. He has pioneered work in organizational culture and leadership, process consulting, and career development. Ed's contributions date back to the early 1960s and continue with the recent publication of *Organizational Culture and Leadership* (2017), *Humble Leadership (2018)*, and *The Corporate Culture Survival Guide* (2019), all co-authored with his son Peter A. Schein.

* This contribution is based on a memorandum I passed out at the beginning of three-day retreat for executive management. I suggest that you ask yourself these very questions and reflect accordingly.

THE DANGER OF ALTERNATIVE DATA

PETER SCHEIN

We have become so good at collecting data, and now we are mining massive data sets to create new derivative and 'predictive' data. It's math; it's great. It is also just as manipulable as it is great.

I recently heard an argument citing a credible medical study that concluded something to the effect that surgical masks may be 30% less effective than previously thought at stopping the transmission of virus-carrying droplets. Perhaps, but the argument, raising technical doubt about surgical masks in COVID-19 hospital use, was introduced to sow doubt about the efficacy of wearing facial coverings in public. This supposed counter-indicating data, manipulated out of context, itself virulently shifts attention to distracting content so as to muddy our perception of the broader context.

And what is the broader contextual observation?

Huge densely-populated metropolitan areas (such as Shanghai, Seoul, Tokyo) have managed to avoid the infection rates that US cities like New York City and Los Angeles have experienced. Let's just suppose in March 2020 that at least 60% of Shanghai residents always wore masks in public, whereas under 30% of NYC residents always wore masks in public. Deep down, we understand the context that we observe, that the wearing of facial coverings of any kind would of course have a demonstrable impact, and the verifiable effectiveness of surgical masks is in fact a separate (distracting) issue. Now, if ~ 90% of us took this social and visible protective measure ~ 90% of the time, assuming it is not a major encumbrance to our life and work, imagine how quickly we could alter the pandemic's trajectory. In fact, we don't have to imagine – we've seen it happen in big cities (just not in the US). Technically verified, perhaps not, but it's there in front of us. Why should we doubt this shared context?

Sowing doubt is very powerful, because 'conclusive' proof is generally only believed by the side that wants to believe it. The 'but what about' arguments that a vocal minority uses to distract and obscure can be more than powerful enough to get the unconvinced to not wear masks and endanger everyone else. When the 'this data shows' declaration builds on the 'what about' claims, we can end up in a bad place, *the doubt stalemate* that freezes conversation, stifles learning, and halts forward progress. This place becomes the worst place when believers and doubters self-righteously refuse to engage in the dialogue needed to reconcile realities (facts and context).

Instead, we should start by being willing to share and acknowledge what we observe, what we think is really going on, without dismissing it as unverified, and consciously engage in sense-making with each other. We are actually making progress when we accept that 'some other way' seems more effective, even if it runs counter to deeply-held beliefs, norms and contrived alternative data. Isn't this common sense? And yet we haven't done it with climate change either. Perhaps this is a lesson we can all take from COVID-19.

Peter Schein is co-founder of the Organizational Culture and Leadership Institute (OCLI.org) and provides counsel to senior management on strategy and OD challenges worldwide. He is co-author of *The Corporate Culture Survival Guide* and *Humble Leadership*. Schein's work draws on industry experience at Apple, SGI, Sun Microsystems and numerous start-ups.

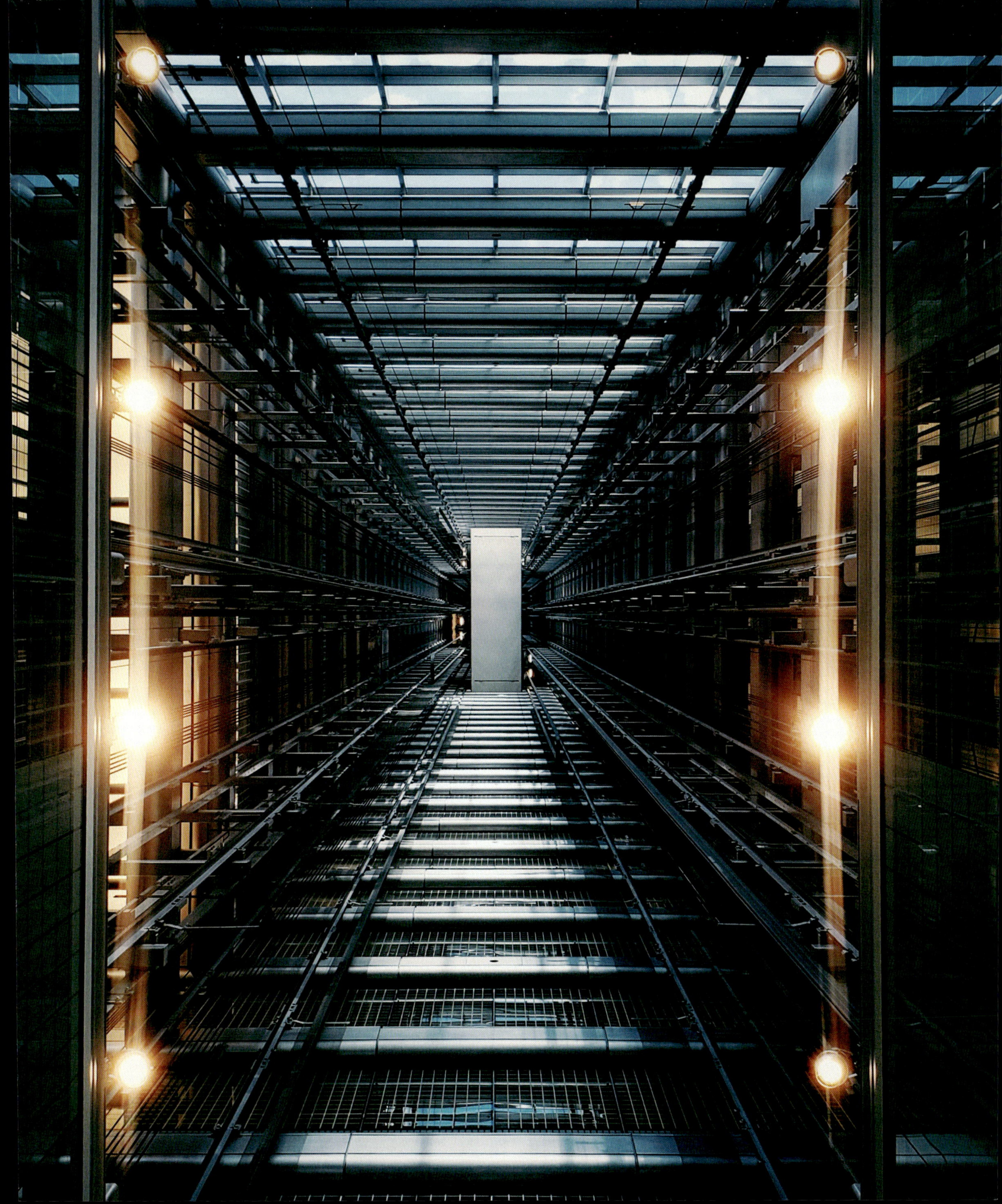

SAFETY FOR THE FUTURE

DAVIDE SCOTTI

Once upon a time, change used to be gradual. You could see it coming, analyze it and even embrace it. Not anymore. In our hyper- connected world, change is exponential and things move incredibly fast. Things we could only imagine yesterday are real today: collaborative robots, artificial intelligence, the Internet of Things, self-driving vehicles, bionics, virtual and augmented reality, wearable technology, big data, 3D and 4D printing ... These innovations amplify one another, creating a perfect storm of change.

What impact will technology have on the way we work and live? What challenges will it pose for health and safety? If robotics and automation may mean people are less likely to work in dangerous environments, the digital workplace will have social, psychological and organizational consequences. Think about the continuous monitoring of workers, the management of work using algorithms, the increase in ergonomic risks caused by new human-machine interfaces or by the online and mobile work. Let's make sure this relentless process of innovation doesn't overwhelm us. *Let's take the lead.*

How? By staying human. In the vortex of ever-changing technology, human-only traits like creativity, imagination, intuition, emotion and ethics will be the one constant. Values like honesty, justice, cooperation, feelings of friendship, trust and love, the search for happiness and the search for safety remain human-centric. All of these things illustrate the endless potential of human life; a picture that machines will never draw. Technology will certainly lend us a hand, but it will never replace the human heart.

We need to get to know technology without being afraid of it, whether that technology is equipment and applications or the landscape of our new digital workplaces. Only then will we make the best use of it and be able to best navigate it to benefit our health and safety. As the futurist Gerd Leonhard said, "The future is in technology, but the bigger future lies in transcending it."

Davide Scotti is an author, speaker and expert in cultural change in the health and safety field. He is the Head of HSE Culture, Communication and Training at Saipem, and since 2010 general secretary of the Leadership in Health & Safety Foundation.

CONNECTING THE DOTS

KATHY A. SEABROOK

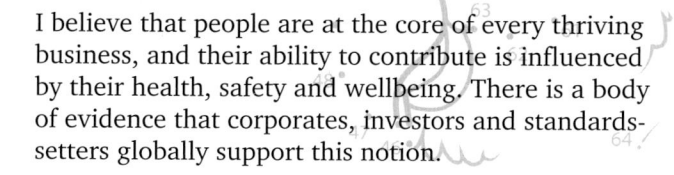

I believe that people are at the core of every thriving business, and their ability to contribute is influenced by their health, safety and wellbeing. There is a body of evidence that corporates, investors and standards-setters globally support this notion.

A thriving business is one that proactively anticipates and assesses risk to both employees and business operations. Strategy, planning and actions – as well as transparency – create value. When a company proactively anticipates and assesses risks to mitigate potential high-consequence incidents, they are prepared in the event an incident occurs. Robust crisis management plans can be executed to bring the company back to a pre- incident performance level or better. That is resilience.

The world is shifting. For many companies, COVID-19 has shined a light on the value of people working in companies and organizations, from frontline healthcare workers to those retooling to make ventilators. It took innovation. It took adaptive people. What COVID-19 has taught us is that companies that had robust risk management planning and a proactive mindset of anticipating and assessing potential threats to their essential people were more resilient post-pandemic. Workers create value for stakeholders and their safety, health and wellbeing is imbedded in this value.

Traditionally, workers were considered a cost to a company, but this premise is being challenged. Now, stakeholders, specifically the investment community, want more transparency for better decision-making on stock ownership. Customers, suppliers, communities, workers and other stakeholders are interested in the workforce's sustainability. The investment community wants to know how well a company is managing risk, specifically to their workforce, also known as human capital. A healthy and safe workforce creates value in the eyes of investors and forward-thinking corporates who see their people as also providing a competitive advantage.

Alignment on purpose and value of workers throughout the company is key to sustainable corporate performance. Operationally, the safety and health of your workforce is imbedded in policies, processes and institutional memory. A worker's discretionary energy drives value; they're more likely to contribute discretionary energy and actively align with their company's purpose when they are engaged and well at work. What does this look like in action? A worker staying late to finish a client project on schedule or working into the next shift to get production back on line following unscheduled downtime due to equipment failure.

Know the business environment in which you operate. Know the risks to the business, operationally, commercially and to the workforce. Managing risks to your workforce influences corporate performance and resilience. It requires collaborative, cross-functional preplanning and good leaders who communicate clearly and concisely at exactly the right time to cut through the ambiguity of the business environment. Be that leader and influence. Ask the right questions: Does your company know its potential high-consequence risks? Would your company be resilient in the event of a low-probability high-consequence incident (such as COVID-19)?

If not, why not? What systematic mechanisms are in place to identify, assess and manage potential risk to your people and the business? What are your next steps?

Kathy A. Seabrook, CEO of Global Solutions, Inc., is a futurist and influencer and works across industry sectors leveraging OSH for operational, commercial and safety/health excellence resulting in sustainable, resilient corporate performance. She is Chair of the Center for Safety & Health Sustainability and a Fellow of IOSH and ASSP.

> I KEEP SIX HONEST SERVING-MEN (THEY TAUGHT ME ALL I KNEW). THEIR NAMES ARE WHAT AND WHY AND WHEN, AND HOW AND WHERE AND WHO.
>
> RUDYARD KIPLING

CURIOSITY DID NOT KILL THE CAT

ANDREW SHARMAN

Most leaders simply tell workers *what* to do. Some tell them *how* to do stuff. Very few leaders really take the time to explain *why* something is important.

Over the last 25 years of working in safety, I can't help but wonder how many leaders really understand *why* safety is important – to their organizations, their people, to themselves. Sure, I hear many asserting *"Safety first!"* or offering that *"Safety is our top priority!"*, though rarely do I see this being truly lived in practice.

Are leaders focusing on the wrong thing when they communicate? As management guru Stephen Covey said: *"Seek first to understand, then to be understood."*

When it comes to understanding safety, many leaders ask questions that are a waste of their breath. I've heard them - a million times now - at the start of meetings or on *Safety Walks* as the leader sets off to do their once-a-week / once-a-month / once-in-a-blue-moon tour of the shop floor:

"Is everything safe?"

"Any safety issues here?"

"Do you have any health and safety suggestions?"

Rarely, if ever, do these questions produce meaningful dialogue as workers respond with a quiet nod or shrug, and the leader walks on.

Do people ask these pointless questions because they believe that old proverb, that *"curiosity killed the cat"*? Maybe it's more than that…

In her studies of leadership, American author Brené Brown suggests that: *"Curiosity builds connection… Connection gives meaning to our lives… [and] Connection is why we're here."* Brown argues that many leaders lack the courage to be truly curious.

Psychologists define curiosity as *'wanting to know'*. That's it. And this simple definition fits well with our common sense, doesn't it? In fact, *'wanting to know'* implies a quest or a search for information.

Curiosity didn't kill any cats.

And it doesn't like rules and systems. It rejects procedures and process. Curiosity loves meandering, diversions and impulsive left turns. Like cats, curiosity is hungry: the more you feed it, the more it wants.

Science starts with curiosity. So does love, friendship, fun, conversation… and, I'd argue, so does safety, too.

What if your Management Safety Walks were a *'search for information'*? How about instead of making inspections and walkarounds you engaged in *'Curious Conversations'* with your people?

Curiosity isn't just a way of understanding the world; it's a way of changing it, too. So, in this brave new world, why is safety important to you? What is it that you want to know about safety in your organization? And which questions do you need to ask in order to find the answers?

Prof Dr Andrew Sharman is Chairman of the Board of the *Institute of Leadership & Management* and President of the *Institution of Occupational Safety & Health* ('IOSH'). He's the author of 9 books on safety culture, leadership, and wellbeing and consults globally on these topics. He loves great questions.

THE 7 FOES

STEVEN SHORROCK

Health and safety professionals have two fundamental activities: understanding and intervention. These activities proceed in various ways. There may be understanding without intervention, intervention without understanding, or any sequence or cycle of these activities. Unfortunately, there are more ways to get things wrong than right, and our OSH efforts can become counter-productive. Here, I'll outline the seven foes of intervention.

1. **Haste**
   When responding to an unwanted event, the need to reduce anxiety associated with uncertainty brings an 'urge for urgency'. It often results in a premature choice of intervention, built on false assumptions about the problem and the evolving system in which it exists (or existed).

2. **Over-reaction**
   A single unwanted event, set against hundreds of thousands of successes, can trigger a system-wide change that makes work harder for many stakeholders, and perhaps riskier. When overreaction and haste are combined, efficiency is favoured over thoroughness, and critical understanding is missing. Unforeseen secondary problems may well be worse than the original one. The result can involve large compensatory adjustments.

3. **Component focus**
   System safety concerns interactions between micro, meso and macro aspects of socio-technical systems – human, social, organizational, regulatory, political, technical, economic, procedural, informational and temporal. Everything is connected to and influences something. The system does something that no component can. But organizational understanding and intervention is often at the level of components. Acting on components blindsides organizations to interactions and system-wide patterns, creating unintended consequences.

4. **Over-proceduralization**
   Work-as-prescribed (rules, procedures, regulations) is necessary to guide work-as-done. But work can rarely be completely prescribed. Work-as-done takes work-as-prescribed as a framework, adjusting and adapting to situations in a dynamic and creative way. To resolve anxiety, nailing down more details is often a favoured intervention strategy. The result is more pressure and fewer degrees of freedom for necessary adjustments.

5. **Scape-goating**
   Blame is a natural tendency following unwanted events. Assigning moral responsibility is – in some cases – necessary, and fundamental to the rule of law, especially regarding intentional harm. But scapegoating singles out and mistreats a person or group for unmerited blame. The result may satisfy outrage or displace responsibility, without solving a wider or deeper problem, leaving the system vulnerable to similar patterns of dysfunction – a moral and practical problem.

6. **'Zero' thinking and targetology**
   This foe involves conflating a measurement and a goal. With 'never' or 'zero' thinking, the implication is that there can be zero harm/accidents, while performance targets or limits may refer to a criterion that must be reached or must not be breached, often with consequences for failure. These blunt interventions often misunderstand the nature of people, systems, accidents, and measurement.

7. **Campaigns**
   Campaigns are a favoured top-down means of change. Unless the activity helps to understand the messy reality of work-as-done in the context of the system as a whole, the effects often wear off shortly after the campaign ends. Staff know this dynamic well; it has been done to and for them many times.

Heed these foes; live and work more safely.

Dr Steven Shorrock is a chartered psychologist and chartered ergonomist and human factors specialist. Since 1997, he has worked on human factors and safety projects internationally. He works at EUROCONTROL and is Editor-in-Chief of *HindSight* magazine. He blogs at www.humanisticsystems.com and co-edited *Human Factors and Ergonomics in Practice*.

THE CLOSET
MARK SIMMONDS

The stress has been building up during the last month or so. Nothing catastrophic, but persistent none the less. I am beginning to sleep fitfully, resorting to a melatonin tablet 30 minutes before bedtime. I am not eating well, socializing less and less, worrying more and more.

The situation is exacerbated by the fact that I just don't feel I can share my dark secret with anybody but my wife. Coming clean at work doesn't seem to be an option. What would my boss, Martin, say? It would be the end of my career, surely.

The end came swiftly. It was Monday afternoon when Martin called me into his office: "Mark, the team and I have noticed that you have not been yourself lately. You've lost your usual spark, your sense of humour. Is everything okay?"

My reaction to Martin's question was pretty immediate. "Yes, I am absolutely fine, thanks. There are just a couple of tricky things going on at home, that's all." My boss stared at me for a couple of long seconds in that "I don't believe you are telling me the truth" kind of way.

"Are you sure? We have all had a lot on our plates during the last few weeks. Even I am beginning to feel the heat." It was at that point that I simply broke down in floods of tears. After the sobbing stopped, I revealed it all, leaving nothing out. Martin listened attentively, sympathetically, kindly.

After 15 minutes of pouring my heart out, my boss apologized but said that he had an urgent meeting that he had to attend. However, he suggested that we met up for a drink after work at the local pub to talk things through. Of course, I accepted, but I was sure that after some initial sympathetic chit chat, this conversation was going to revolve around my career choices.

At 7 pm, I walked into the pub and was absolutely horrified to see Martin sitting down with Mary and Jim, my two work colleagues, drinks on the table, laughing and joking. Martin ushered me over, sat me down and went to the bar to buy me a drink. He returned, drink in hand and sat down. He looked at me and then announced out loud:

"Mark, welcome to the Nutters Club! A club for people who struggle with different mental issues but who have decided to come out of their respective closets. Our motto is 'Better out than in'. You are now officially a lifetime member. Cheers everyone!"

So, what's the moral of the story? In the workplace, it's senior management who hold the key to mental health. By understanding it, normalizing it, accepting it and removing the stigma, the irony is that they will also create a working environment which is *more* rather than less productive. It will likely be healthier one. It will certainly be a happier one. And that must be a good thing.

Mark Simmonds trains major organizations in creativity, insight and innovation. He talks about his experiences with mental health both as a sufferer and as a caregiver in his book *Breakdown and Repair*.

THE PSYCHOLOGICAL VOLCANO

KARL SIMONS

I have said for many years, "It's not the psychological condition that prevents a person from thriving at work, it's the environment in which they are placed." This simple message explained to managers that they can make adjustments that aid people working under their leadership to have more fulfilling and productive work lives. This is true even for workers with a clinically diagnosed illness.

Front-line operational leaders should take time to understand the link between the psychology of human behaviour and its effect on individuals' ability to suppress their feelings. Individuals develop coping mechanisms to enable them to overcome adversity whenever it presents itself at work, often described as "pulling on the corporate veil." This usually means a bubbling swamp of festering issues that go unnoticed by the management or peers who may also be experiencing similar issues in the workplace. Ultimately, this then leads to the resulting absence occurring once it becomes too much for a person to cope with.

It can be explained as a 'psychological volcano' that, in simple terms, shows the connection between the internal make up of an active volcano and a person:

- A volcano is made up of layers of rock and ash that, over the years, have enabled it to suppress the bubbling magma at its base, but on occasion the pressure from the magma builds to a level where an eruption occurs.
- A person psychologically, over time, builds levels of resilience and coping mechanisms based on experience that enables him/her to suppress the boiling frustration they feel in the face of adversity, however on occasion the pressure becomes to great and leads to them speaking out or acting in anger.

Removing the pressure through removing the stressors that are causing it – be it improving the working environment culture or the lifestyle issues – should be considered an organizational goal for their people and teams.

Over the years, I've listened to a lot of academics telling people what you can do and what you should do, but very few practitioners are saying: "Here's what I've done, and here are the outcomes achieved as a result!" OSH practitioners have embedded safety management controls into work practices for many years, making reasonable adjustments to support people, yet regrettably few have applied the same thinking to their systems and controls for psychological health management. It is clear to me that no one single initiative can achieve a cultural shift; you need to be relentless over consecutive years. A case study can be seen of a five-year mental health program that was implemented, through waves of psychological health initiatives, into the United Kingdom's largest water company Thames Water between 2013-2018. This yielded an 80% reduction in work-related Illness in that period.

As leaders, we can have a concrete positive effect on workers' psychological health by changing a few factors. Otherwise, we risk the volcano erupting.

Karl Simons is Chief Health, Safety & Wellbeing Officer at Thames Water. He served in HM Armed Forces and worked globally in safety-critical industries. He's an adviser to the UK government, a lecturer at the University of Cambridge, and a Non-Executive Director for 'Water & Sanitation for the Urban Poor.

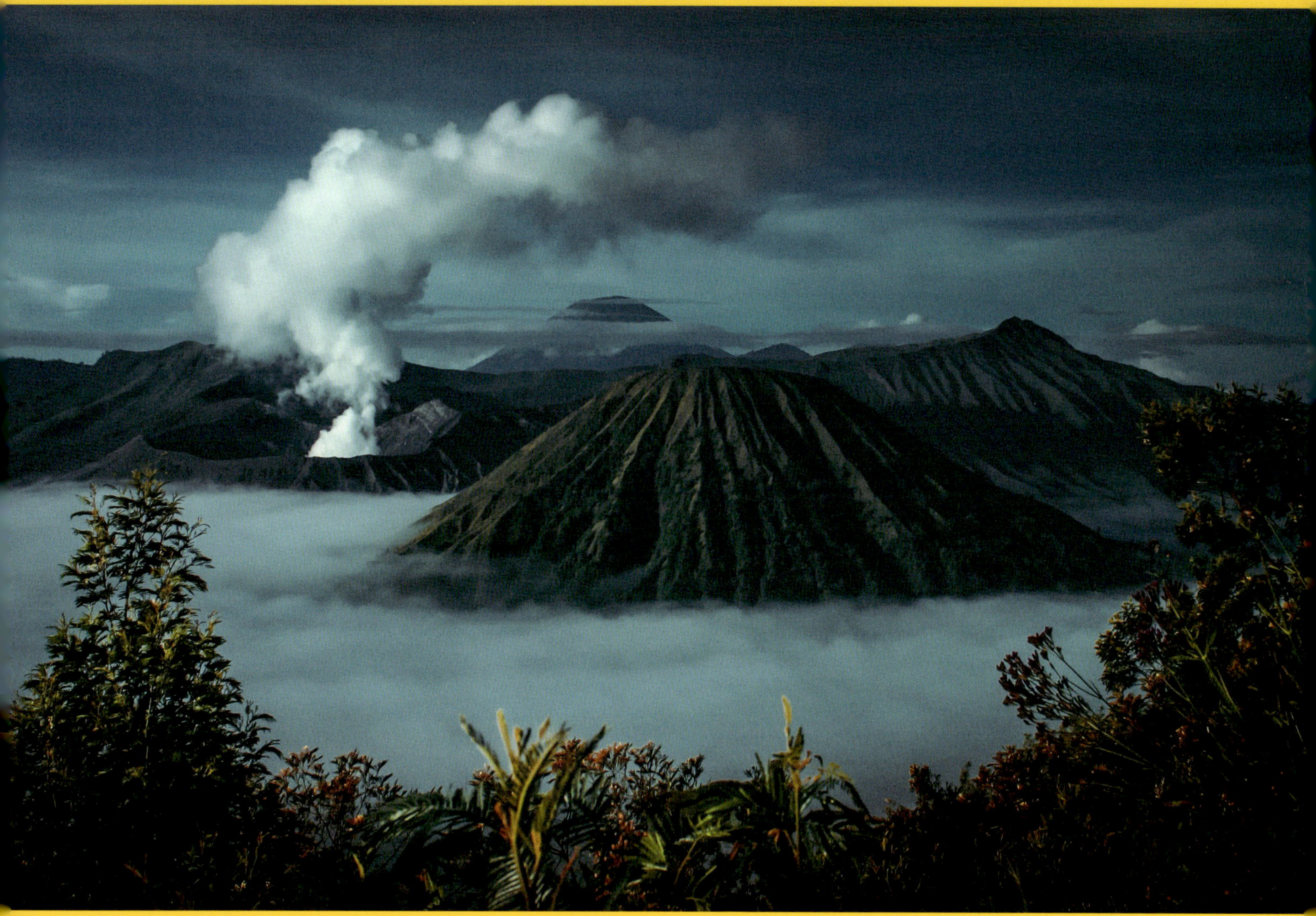

MAYBE IF OUR LEADERS HAVE THE COURAGE TO SHARE THE NAMES AND PHOTOS OF PEOPLE WHO HAVE DIED AT WORK, RATHER THAN ONLY FATALITY STATISTICS, OUR WORLD WILL BEGIN TO BECOME SAFER, SINGLE LIFE BY SINGLE LIFE

ONE PERCENT IS 1 IN 100

PAUL SLOVIC

The high value we place upon individual lives is demonstrated by the strenuous actions we often take to protect single individuals. Those of us who study the value of human lives have a name for the importance of saving one life. We call it 'the singularity effect.'

Opposite singularity is a response, 'psychic numbing,' which happens when the threat is great – many individuals are at risk – and the information about the threat to their lives is communicated to us as statistics. What we have learned from experience as well as from scientific research, is that it is hard to appreciate the humanity that large numbers represent. As someone aptly said, "statistics are human beings with the tears dried off". Another similarly observed "One man's death is a tragedy; a million deaths is a statistic". These oft-cited expressions convey a harsh insight into what psychologists have described as a flawed and deadly 'arithmetic of compassion'. In contrast to individual deaths, which are emotionally wrenching, numerical data are numbing, in that they fail to spark the feelings and emotions that are needed to motivate us to act.

This bias in our humanitarian accounting has been documented in numerous psychological experiments on life-saving behavior, showing that our intuitive feelings—which we trust to guide us in making all manner of decisions—are innumerate. As the number of lives at risk increases, psychic numbing begins to desensitize us. A single life at risk feels less valuable to protect if it is part of a larger tragedy, with many lives endangered. You probably won't feel any more concerned learning about a threat to 88 lives than you feel about a threat to 87 lives, unless you pause, do the math, and realize that there is one additional, and valuable, life at risk.

Research on the feelings associated with risk has taught us that the way information is framed can strongly affect our appreciation of its meaning. One percent means 1 in 100, but our minds react differently to these two logically identical frames. When we read "one percent saving of lives," we may think of a small number. When we read "1 out of every 100", we likely think of one life and how important it is.

Maybe if our leaders have the courage to share the names and photos of people who have died at work, rather than only fatality statistics, our world will begin to become safer, single life by single life.

Dr Paul Slovic is professor of psychology at the University of Oregon and founder and President of Decision Research. He studies human judgment, decision making and the psychology of risk. His most recent research examines 'psychic numbing' and the failure to respond appropriately to mass human tragedies.

ARE YOU READY FOR THE NEW NORMAL?

CAROLE SPIERS

On June 1, 2020, British schools began re-opening and retail stores prepared to welcome shoppers as the government lockdown began to ease. Whilst the news of reopening made some jump for joy, others experienced feelings of anxiety as they prepared to resume their lives in a post-lockdown world.

Anxiety can affect many people and be triggered by various events. One form of anxiety labelled 're-entry anxiety' is a type of stress caused by the inability to adapt to previous routines or the desire not to return to these routines. This might be the case following a car accident, where the victim is unwilling to get back into their car.

It is also important to recognize that the routines we once followed in our pre-lockdown lives weren't restricted by social distancing measures and fears of contracting a virus. Whether or not you caught the virus yourself, many can identify with this anxiety around the idea of returning to 'normal,' after a prolonged period where normality was denied.

These anxieties may be even greater, as none of us have lived through a pandemic before. We know that re-entry is a good thing, however, it is important to do this at a pace that is comfortable. Employers should take into account the feelings of their employees when they are permitted to return to the workplace. Many employees will be keen to get back to business as usual, whilst others will prefer to work from the safety of their own homes.

Communication is also critical, so that employees are clear about what will be expected of them going forwards. They should be kept in the loop throughout the entire decision-making process to reduce stress and increase confidence. These are tough times that call for caution and understanding. Leaders ought to demonstrate active listening, care and support.

A leader's job has always included getting the most out of their teams in terms of performance and productivity, however there may be new ways to achieve this goal.

Living through this pandemic has forced us to move out of our comfort zones in order to survive. The learning associated with this needs to be internalized and taken forward as we focus on the future.

Many of us have had the time to pause, reflect and reconsider how we want to live and work.

While many have been eager to resume their previous lives, some say that this time has forced them to stop, think and breathe. It's been a time when we put things aside to go for a walk, to chat with loved ones and to enjoy moments alone. A time when we were grateful for every day.

We can all be guilty of having a short memory, but I do hope going forward that the learning we have gained through this fraught period will stay with us for a long time. These lessons of transparency, support and mindfulness should stay with us, even when the pandemic is in the distant past.

Carole Spiers FISMA, FPSA, MIHPE, is CEO of a leading UK stress management and wellbeing consultancy. Spiers is an international motivational speaker and Executive Coach. She is Chair of the International Stress Management Association UK, founder of International Stress Awareness Week, Fellow and Past President of the Professional Speaking Association, London.

MINDFULNESS AS A SAFETY RESOURCE

CHRISTIANE SPITZMUELLER & PEGGY LINDNER

The risks of working in many industries are well-documented, and workplace accidents and fatalities, as well as process safety incidents, continue to affect millions worldwide every year. We advocate, through our recent work with industry partners and funded by a National Academies of Sciences, Engineering and Medicine grant, to examine how, when and why mindfulness interventions can promote workplace safety.

A large body of interdisciplinary literature shows that mindfulness interventions and exercises can effectively address attentional deficits, enhance worker well-being and result in improved employee outcomes. Through our work, a partnership between researchers and two energy companies, we are aiming to build new evidence-based mindfulness practice tools to enhance safety outcomes.

Specifically, the partnership is examining whether simple exercises designed to increase mindfulness can improve safety. We will test the impact of our mindfulness interventions through work with a high-risk, high- stress population: Offshore workers in the Gulf of Mexico. Participating offshore workers will undergo training as part of a 30-day intervention, consisting of two brief exercises daily. One of the training exercises will be conducted during group meetings as part of the start of a shift, and the second exercise will be done independently by workers in conjunction with a task that is identified as a task with inherent personal or process safety risks. Follow-up employee surveys and company records will be integrated into a database to allow analysis.

The implemented mindfulness exercises teach people to focus on the moment and on specific tasks. Based on research in other industries such as healthcare, mindfulness is expected to be linked to better safety outcomes. In this case, 'mindfulness' isn't a new-age concept, but one designed to signify better awareness of what is happening on the platform. The exercises will be modified from those used for the military and first responders to make sure they match the cultural environment offshore.

In our project, we define success as lowering rates of safety-related incidents, both for process safety – things like equipment downtime– as well as personal safety and enhanced worker health and wellbeing. Participants will be surveyed to determine if health and sleep has improved, benchmarked against workers who did not participate in the exercises.

Altogether, we expect to add mindfulness interventions to the repertoire of safety practitioners to show how, when and why they can help augment safety culture and employee health and wellbeing.

Dr Christiane Spitzmueller is Professor of Psychology at the University of Houston (UH), where she conducts research and teaches courses in Occupational Health Psychology and training and development. Her research has been published in top HR journals in her field, and her research has been cited more than 3,000 times.

BE THE HUMMINGBIRD

MALCOLM STAVES

Before starting, I want you to know that I attribute anything I have achieved to the teams I have had the privilege to work with. So, my advice follows in line with this experience. I offer seven points that have guided me in my life and work, and I hope they help to guide you as well.

1. *Make sure you have a good, well-balanced Occupational Health & Safety (OSH) leadership team that can advise and support your business.*

Continuous personal development in leadership areas is vital for any professional, as influencing and supporting management is not just about expertise.

2. *Risk management needs to be your foundation, and from there, you can evolve a health and safety culture in order to drive toward excellence and beyond.*

In my first leadership training, I was told I cared *too much* about people and needed to change. I remained true to myself, and I now am convinced that putting people and their wellbeing at the center of everything we do is critical for agile and resilient organizations. Gone are the top-down military-style ways of management, which leads me to:

3. *Put people in the centre of what you do, be passionate and lead with the heart.*

When it comes to developing health and safety culture within an organization, it is important to have people you can count on and have a passion for health and safety at all levels.

4. *It is your passion that will light the passion in others, but you must always 'walk the talk', even at challenging moments.*
5. *Surround yourself with business leaders that are passionate for health and safety efforts.*

When it comes to OSH in practice, I am a great advocate of driving the pillars of risk management, towards excellence and beyond, driving towards an interdependent OSH culture.

6. *Give a sense of "Why safety matters" – don't look to the organization for the answer. Be bold; make a difference both inside and outside the organization.*

Ultimately, I'm guided by three animals. First, the elephant, who is well-organized, unwavering and cares for the family and the herd. Second, it's the canary, who represents looking relentlessly for low-level signals of risk (and acting on them). Finally, it's the hummingbird, who signifies that progress is made in small steps. All of these all together make a big difference.

Be the Hummingbird (you may need to look up *The Legend of the Hummingbird*).

Think big, act small, be bold. Your OSH culture will evolve, and you'll create health and safety ambassadors of all your employees.

I hope that you are never the one to have to say, "I could have saved someone's life today but I chose to look the other way." Let's continue to "be the hummingbird" and make small changes that will have big reverberations throughout our organizations and our lives.

Malcolm Staves is a Fellow of the Institution of Chemical Engineers. He worked in various EHS positions before joining L'Oréal as Group Health & Safety Director. Staves is a member of the Center for Safety and Health Sustainability Advisory Board and a Capital Coalitions Occupational Health & Safety expert partner.

THINK BIG,
ACT SMALL,
BE BOLD

HOW ARE YOU TODAY?

A QUESTION OF HEALTH

ROB STEPHENSON

What if I were to ask you a question?

What if I were to tell you that we are asked this question 10, 20, 30 times per day?

And what if I were to tell you that we very rarely answer this question honestly?

But what if I were to help you answer it honestly? And by doing so it could change your life and the lives of your employees, family members and contacts?

The question is "How are you?" and is usually met with a ritualistic response of "I'm good"; "I'm fine"; "I'm alright"; "I'm okay" or, my favourite British response: "Not too bad." We can do better.

So, yes, we are focusing on the much-neglected health element of the health and safety agenda, and specifically, the mental health of the humans we interact with. But we are not going to talk about mental health here. Why? Because of the stigma. It gets in the way. Yes, we must smash that stigma, and leaders have a critical role to play here by sharing their stories of mental ill-health, but I also think that 'mental health' needs a rebrand. So let's talk about 'form' instead, answering the question "How are you today?" with a score, out of ten.

At the time of writing, my FormScore is a 7/10 – I am on good form. Yesterday I was a 5/10 and on poor form. Like our physical fitness, our mental form is always moving. Adding "today" to the question takes away the ritual and encourages one to pause and reflect. Giving an answer out of ten makes people feel more comfortable sharing.

Why should you care?

We lose a life to suicide globally every 40 seconds.

Only 17% of people in a given population are thriving at any one time. (Keyes, Huppert 2005)

The cost of mental ill-health to UK employers is estimated at £45 billion a year. (Deloitte 2020)

Focusing on helping people manage their form is one of the biggest performance gains we can make. And it will save lives.

So, start with yourself. How have you slept? Have you exercised regularly? How has your nutrition been? Are you balancing stress with moments of recovery? Are you worried or relaxed about your finances? Do you have a strong sense of purpose? Do you find simple tasks and decisions easy or difficult? What are your energy levels like? Are you helping others? How connected are you to friends and family?

Notice how you are today, out of ten. And then connect with a colleague; a loved one or a friend and share your score. Ask them the question "How are you today?" and watch the magic happen.

Rob Stephenson is an international speaker, campaigner and consultant on a mission to help create happier, healthier and higher performing workplaces. Stephenson experiences bipolar disorder personally. He is the founder of InsideOut and the CEO of Form, a movement helping employees authentically answer the question "How are you today?"

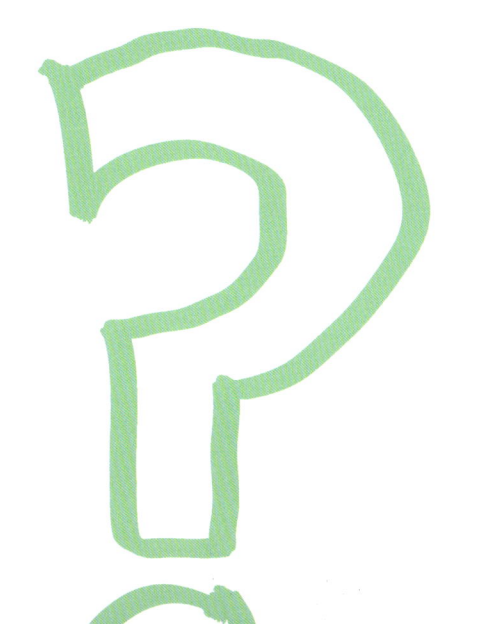

TARGETS MATTER

NEAL STONE

The importance of targets cannot be over-stated. In June 2000, the UK government's Department of Environment, Transport and Regions, in conjunction with HSE, published a ground-breaking strategy, *Revitalising Health and Safety*, "to give a new impetus to health and safety at work." The strategy recognized that while substantial improvements had been made, much still needed to be done to prevent the tragedy of workplace fatalities and injuries, work-related diseases and ill health.

For the first time, the UK government and HSE set precise targets for reducing:

- The number of days lost due to work-related injury and ill health
- The incidence of fatal and major injury
- The incidence of work-related ill health cases

For some, the targets were an inconvenience. A lack of will both by politicians and officials resulted in them being side-lined and then binned. But despite their brief existence, the impact that measurable targets had was considerable. They did wonders across many industrial sectors – helping to concentrate minds and improve the control of risk.

Targets alone are not enough. Key to success in reducing injuries and ill health is a sound regulatory system, competent people to ensure workplace risks are effectively managed, strong visible leadership, and an actively involved workforce. Without these, effective control of workplace risks and sustained improvements will not be achieved.

There is no need for more evidence or a revolutionary approach in our thinking concerning workforce engagement. The evidence from many studies carried out over the last twenty years consistently point to a number of factors which contribute to effective workforce engagement including:

- Lifting the veil of complexity that is perceived to surround health and safety
- Helping individuals better understand the risks the workplace poses and the contribution they can make though their own behaviour
- Organizations demonstrating through words and behaviours the value they attach to having an engaged participative workforce
- Ensuring cost is not used to impede good practice.

What the COVID-19 pandemic has demonstrated is the importance of having goals that the public and practitioners identify with and ones that have a profound influence on behaviour. Without these goals and general public acceptance of the need to comply with unprecedented rules, the rate of contagion and resulting deaths would have been so much higher.

The management of workplace health and safety must look long and hard at the levers that were successfully pulled to build public awareness and understanding of what significant risks the COVID-19 pandemic posed, the targets that needed to be achieved and the behaviours needed to contain the risk of contagion. Then, we can apply these lessons to our workplaces going forward.

Neal Stone LLB MSc MPhil is currently Health and Safety Policy Adviser for the communications agency McOnie. From 2008-2016, Neal worked for the British Safety Council, ultimately as Deputy Chief Executive and Policy Director. Previously, Neal served as a policy adviser at the Health and Safety Executive from 1992-2008.

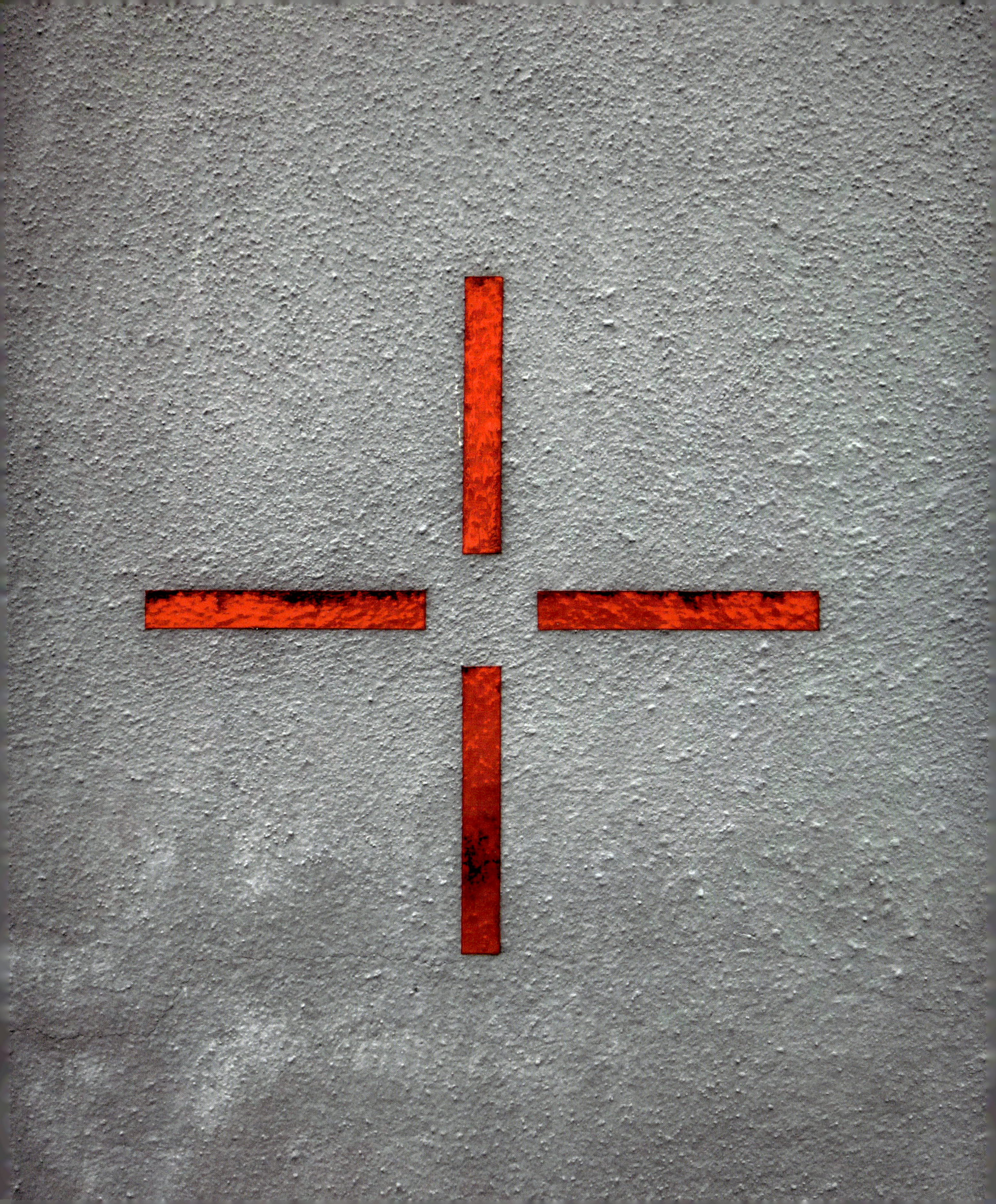

GOOD MORNING
ENJOY!
AT THE END OF EVERY SHIFT ONLY 1 OUTCOME MATTERS
AVOID THE ZERO ACCIDENT TRAP!
EVERYONE HOME SAFE EVERY DAY
FROM PREVENTING ACCIDENTS
OUR JOINT RESPONSIBILITY!!
TO CREATING SAFETY
SO GLAD YOU'RE HOME!
WE OWE IT TO THE ONES WE LOVE
DO YOU HAVE A CULTURE SAFE ENOUGH TO SHARE ALL STORIES
EVERY STORY IS CRUCIAL
EXTREMELY HARD
DIFFICULT
EASY
THE UGLY
THE BAD
THE GOOD
let's talk
SAFETY
STOP
START
IMPROVE
KEEP

EVERYONE, EVERY DAY

EDWIN STOOP

Edwin Stoop is the founder and Chief Creative Officer of Sketching Maniacs (www.sketchingmaniacs.com). Edwin is a strong believer of the power of visual stories as a way of boosting engagement, connection, understanding and movement. Welcome to the other side!

GIVE ME CURIOSITY OVER EXPERIENCE

BRIAN SUTTON

As leaders and managers, we are conditioned to value experience. If we are filling an important position and have to choose between a candidate with 10 years experience gained in three companies on two continents and a candidate with one year's experience in one organization, it's a no-brainer – we go with the first candidate. We value experience so much that sometimes we have gathered it together in a pile and put it on a pedestal to be admired. We are so pleased with ourselves that we invented a special name for this: 'best practice'. Before long, we had so many pedestals and so many people wishing to admire what was on them that we had to create special buildings to put them in, museums. Museums are fine places, but I don't want to work in one.

So what is experience? Experience is the personal store of memories of incidents and situations that allows us to recognize the familiar in the unfamiliar. When faced with a unique situation, the experienced operative can say, "Ah, this looks a little like X, and when I see X I do Y." This is great if you value decisiveness, but it has a fatal flaw. The flaw is that we remember what we did when we saw X, rather than what happened as a result of what we did. The consequence is a common phenomenon – we repeat solutions that fail.

When I am looking for a new hire or a consultant, experience is just the initial filter. What I am looking for is curiosity. Curious people see things that others don't; curious people are more open to being surprised. How often have you said to yourself, "Wow, I didn't expect that", and then just carried on regardless? A curious person recognizes the unexpected and uses it to jump-start the learning process. They immediately ask themselves questions such as: "What did I expect?" "Why did I expect that?" "What is it about my understanding of the world that led me to expect that?" "What theory am I using to come to that conclusion?"

Don't get worried by 'theory'; I am not talking about esoteric academic arguments. In simple terms, a theory is a story we tell ourselves to explain the reality that we are experiencing. Most of these stories are unconscious and unexamined, but they govern how we see and act in the world.

When faced with a unique and surprising situation, the experienced focus on trying to match it to something they have seen before as a short cut to action. Whereas, the curious practitioner asks themselves if today the day they need to start telling themselves a different story about how the world works. While the experienced hate surprises, the curious rush up to surprises and give them a hug. Curious people learn faster and deeper and are full of wonder. They are constantly re-writing their understanding of the world.

Be curious, welcome surprise, be open to wonder and make today the day that you start telling yourself a new story.

Dr Brian Sutton is Professor of Learning Performance at Middlesex University, UK. He works with senior professionals who wish to gain doctoral recognition by researching their own practice in their unique operational situation.

BE CURIOUS

EXPLORE

THE UNITED STATES OF AMERICA
FEDERAL RESERVE NOTE
THIS NOTE IS LEGAL TENDER FOR ALL DEBTS, PUBLIC AND PRIVATE
WASHINGTON, D.C.
ONE DOLLAR
E95597674E
G88969823G
F50088469
D38607515E

OUR MONEY OR YOUR LIFE

DARREN SUTTON

Have you ever offered someone extra money to do something for you? I'm not talking about paying them for work they've done or are about to do but offered them a bit extra to give it their all.

Maybe you wanted them to try especially hard or go that extra mile and you wanted to incentivize them for doing things exactly the way you wanted it done? Or maybe you offered them that little bit extra because you suspected that the task was *intrinsically* difficult or tedious, and they didn't really want to do it.

It's so easy to do isn't it? Whether it's motivating our child to do homework well or to take their medicine, offering rewards just seems to work. This is *extrinsic* motivation, persuading people to do things they don't really want to do by offering them a reward or incentive. It's always felt a bit uncomfortable to me, dirty even!

I remember a school mate offering his older brother half his lunch money to help him with his homework. Why wouldn't the elder brother *intrinsically* want to help him? What kind of relationship must they have? It got worse. As time went on, I noticed the same elder brother was being rewarded for *not* doing something. You guessed it, he was being incentivized to not bully people.

My curiosity surrounding extrinsic motivation extended to other performance domains. A father encouraging his 12-year-old son to score three goals today for a handsome monetary reward. A friend whose mother motivated her with the promise of driving lessons, but only if she got nine straight A grades.

Stop and think for a while about the unintended outcomes *extrinsic* motivation often lead to and the potentially damaging effects on *intrinsic* motivators in the longer term, including the team and the family as a whole.

I often hear incentivization works best in sales. It really doesn't, and often CEOs and FDs of organizations, when they become aware of the malpractice and fraudulent behaviour of their sales teams, are first astonished and then disbelieve of the kind of culture and values that lead to such behaviours.

"How did things get so bad?" they wonder.

So why do we still do this for safety? Well, it's for exactly the same reasons as the brother, the parents and the sales directors, I guess. Because it's easier; it's effective short term, and we don't really care enough to optimize the performance in this area with good leadership principles.

Offering rewards for preventing accidents or reporting near misses says that the leadership is not *intrinsically* driven for safety. It's lazy leadership, and we know extrinsic motivators will erode the the best efforts of others in creating the kind of sustainable excellence in safety performance that our people deserve.

Stop rewarding safety performance. Be a better leader: start spending time not on your bonus budget, but by improving the way you communicate to shift mindsets to *wanting* to work safely rather than bribing people.

Darren Sutton is a performance psychologist who enjoys helping people and their organizations become the very best they can be. He co-developed the world's only fully IOSH accredited Behavioural Safety Leadership Program. He is Senior Partner at RMS Switzerland, improving culture and enabling excellence in multinational corporations around the world.

FOCUS ON DISEASE
JUKKA TAKALA

Work-related Deaths caused by Illness and Injury – *Every year* (ILO 2017)

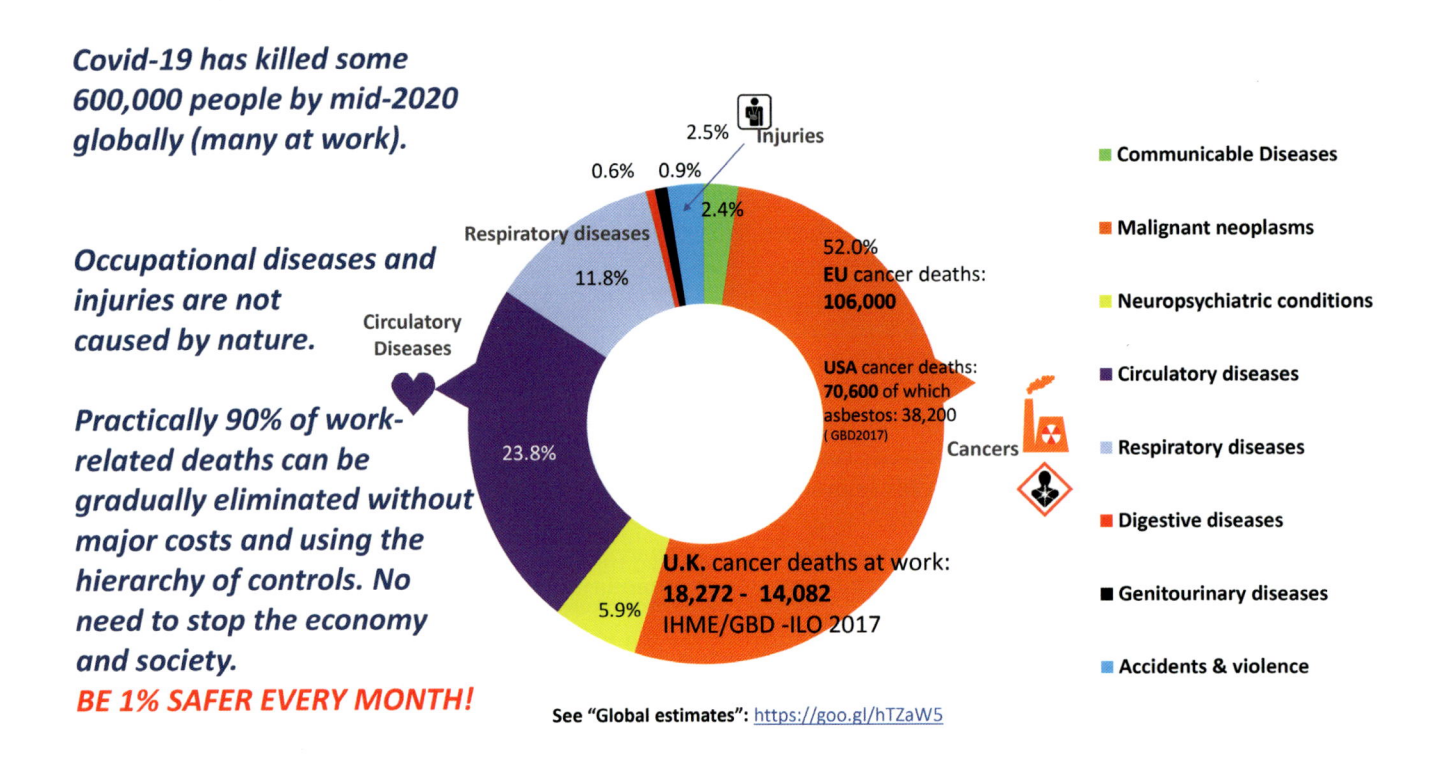

Covid-19 has killed some 600,000 people by mid-2020 globally (many at work).

Occupational diseases and injuries are not caused by nature.

Practically 90% of work-related deaths can be gradually eliminated without major costs and using the hierarchy of controls. No need to stop the economy and society.
BE 1% SAFER EVERY MONTH!

Injuries 2.5%
0.9%
0.6%
2.4%
Respiratory diseases 11.8%
Circulatory Diseases 23.8%
5.9%
52.0% EU cancer deaths: 106,000
USA cancer deaths: 70,600 of which asbestos: 38,200 (GBD2017)
U.K. cancer deaths at work: 18,272 – 14,082
IHME/GBD -ILO 2017
Cancers

- Communicable Diseases
- Malignant neoplasms
- Neuropsychiatric conditions
- Circulatory diseases
- Respiratory diseases
- Digestive diseases
- Genitourinary diseases
- Accidents & violence

See "Global estimates": https://goo.gl/hTZaW5

Professor Dr Jukka Takala is President of the International Commission of Occupational Health with 40+ years of global experience in workplace safety and health in industry and international civil service including Director of the EU-OSHA agency and senior leadership positions in the International Labour Organization and the United Nations.

221

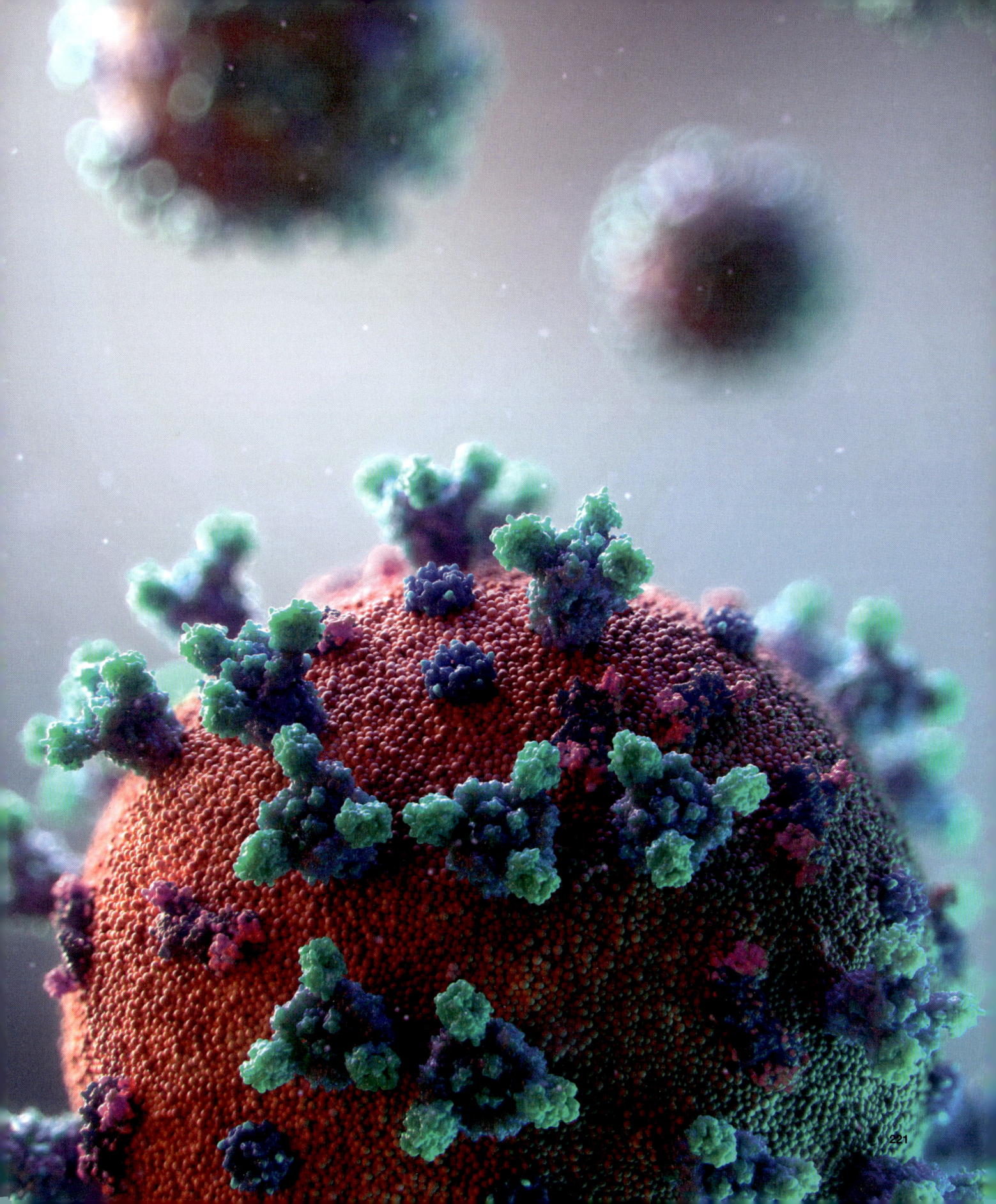

RECOVERING THROUGH COLLABORATION

ANNA-MARIA TEPERI

Despite long-term occupational health and safety measures, sudden and unexpected critical incidents still occur that challenge personnel's ability to cope. At workplaces, less critical incidents, such as near-misses, conflicts and disturbances are more common than severe accidents. These incidents pose a risk to work ability and cause stress reactions, especially if they constantly reoccur or are not properly handled. Personnel at all levels may suffer from critical incident stress, and symptoms may emerge on cognitive, emotional, social and functional levels, reducing overall well-being and the functional capacity needed at work.

However, it is surprising that many organizations still have no systematic methods for jointly handling these critical incidents[1]. A model should therefore be implemented at workplaces to support personnel's resilience, ability to mitigate, cope with, recover and learn from critical incidents. There is a need for a concrete, easy-to-use model for the personnel themselves, and a low threshold for its use at the workplace after the incident. This would avoid having to turn to occupational health services (OHS) or other external professionals.

The Mental Fist Aid (MFA) model, based on the CISM framework[2] involves peer-driven (colleague-to-colleague) support, and includes a structured, small-group discussion within 8–12 hours of the occurrence of the critical incident in order to rapidly reduce and normalize the reactions caused by the event, and to support recovery after it. The MFA has shown valuable user experience in aviation[3] and the public sector[4]. It has helped members of organizations practically handle workplace incidents in everyday work and better handle the subsequent post-incident (stress) reactions at work. It has added value to traditional OSH practices by deepening the participants' understanding and improving their practical skills for managing safety from the human variability perspective, which is regarded as essential in the new safety paradigm[5].

The MFA model is implemented gradually and according to participative work development principles, which are regarded as the most efficient way to improve safety while helping the personnel and management of an organization become the subjects of development activities. It has a positive and significant effect on sense of community, communication and collaboration[6]. The phases of implementation in organizations have been: informing and marketing the model application, nominating and training selected peers to conduct MFA discussions at the workplace immediately after incidents, training supervisors and OSH officials, and trainers' training.

MFA is recommended in any sector or workplace, to commit the personnel themselves to being the subjects of safety. The model has the best result when it is implemented as a strategic choice, through a systematic, long-term process with wide collaboration and empowerment across and within different organizational partners, including management, personnel, HR, occupational safety officials and OHS. It has the potential to enhance organizational and safety culture and leadership, as well as to improve motivation and work satisfaction.

1. Mitchell, J. Critical incident stress management (CISM) – Group crisis intervention. 4th edition. International Critical Incident Stress Foundation, Inc. Ellicott city: 2006. Caine, R.M., Badgasarian, L.T. Early identification and management of critical incident stress. Critical care nurse. 2003;1(23):59-65.
2. Everly, G, Flannery, R.B., Mitchell, J. Critical incident stress management (CISM) – A review of the literature. Aggression and Violent Behavior. 2000;1: 23–40.
3. Leonhardt, J., Vogt, J. Critical incident stress management in aviation. Aldershot: Ashgate; 2006.
4. Teperi AM, Pajala R, Lantto E, Kurki AL. Peer-Based Mental First Aid Following Workplace Incidents: Implementation Evaluation. J Occup Environ Med. 2019;61(8):659 668. doi:10.1097/JOM.0000000000001625
5. Hollnagel, E. Safety-I and Safety-II. The past and future of safety management. Ashgate; 2014.
6. Zwetsloot, G.I.J.M. Evidence of the benefits of a culture of prevention. In: Aaltonen, M. ed. From risks to vision zero proceedings of the international symposium on culture of prevention – Future approaches. Helsinki: Finnish Institute of Occupational Health; 2014:30-35.

Dr Anna-Maria Teperi is Chief Researcher in the Finnish Institute of Occupational Health (FIOH). She is an Adjunct Professor at Tampere University in the field of Human Factors in Safety Management, with 20 years of experience in research and implementing HF in safety-critical fields such as aviation.

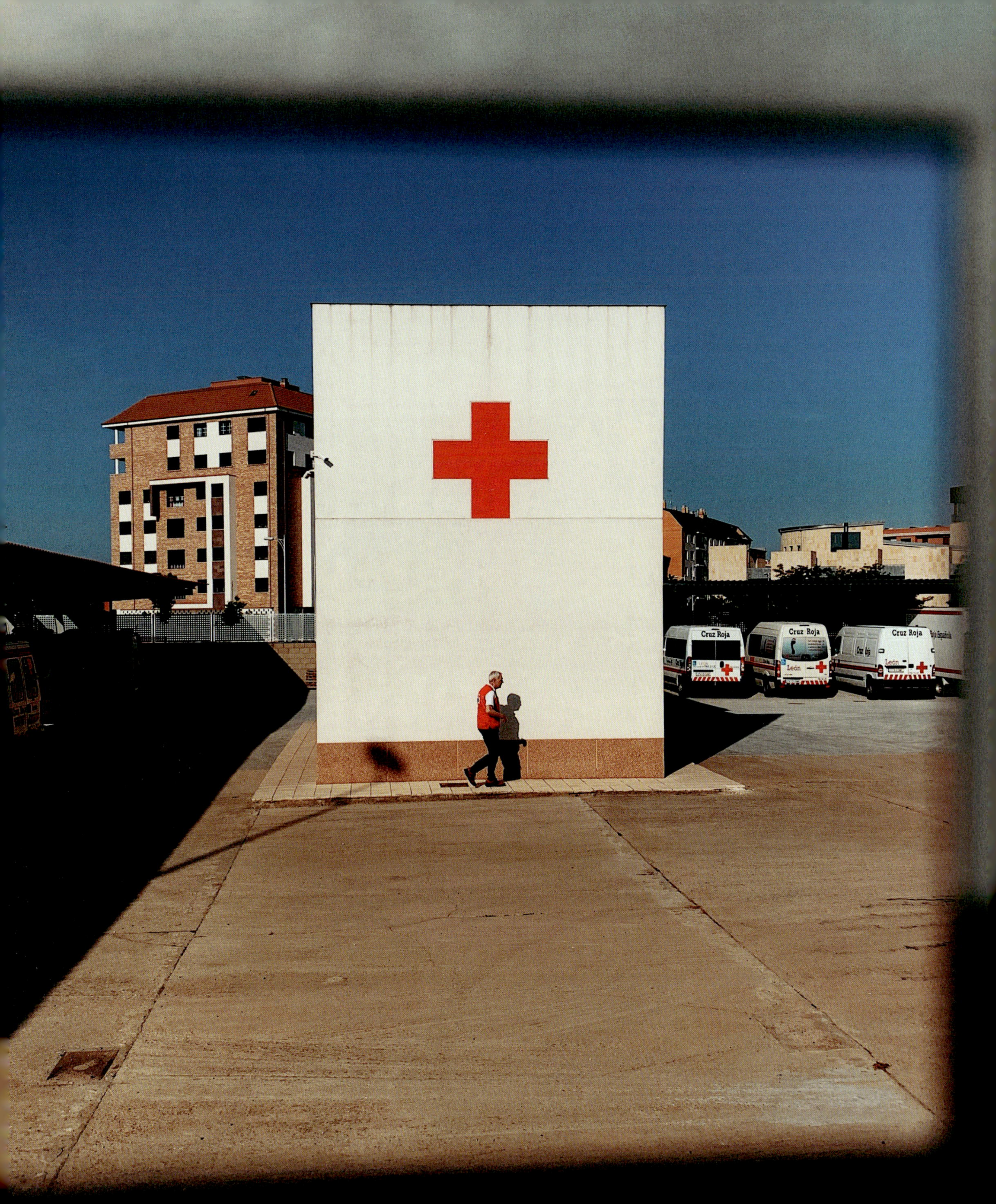
Cruz Roja
Cruz Roja
León
Cruz Roja
León

SMALL BUT MIGHTY

SVEN TIMM

Small businesses are making up the biggest share in the worldwide economy. Globally, they contribute a bigger share in added value than all big enterprises together, but they comparatively perform worse in health and safety at work. These '7 Golden Rules' will guide small business owners to implementing basic preventive measures, resulting in an enhancement of the workers' lives in terms of health, safety and well-being.

Most economic activities worldwide are in businesses small by size, but their importance to both developed and developing economies and societies is indisputable. According to the World Trade Organization, small-and medium-sized enterprises represent over 90% of the business population, 60-70% of employment and 55% of GDP in developed economies. An approach to prevention is predominantly weak or non-existent. Implementation of occupational safety and health measures to protect the working people is essential to making the businesses successful and sustainable in the long term. Now, this is either not considered by the business owner or not the focus. Additionally, small businesses often only have an informal status and are often chronically under-resourced.

'Vision Zero' is the vision of a world without occupational accidents and work-related diseases. Its highest priority is to prevent fatal and serious work accidents and occupational diseases. It's the goal of a comprehensive culture of prevention. The expanded guide "7 Golden Rules for the Health and Safety of Small Businesses" is an important part of the Vision Zero campaign of the International Social Security Association ISSA. The 7GR for small businesses focus on the basic prevention needs of this big share of economic activities. It offers a challenge for small businesses, where the employer is responsible to alone manage the safety, health and well-being of the employees and role model behaviour of self-care. The 7 GR are intended to function as a door opener, attracting small business operators to implement preventive measures in their businesses.

Each of the 7 Golden Rules is addressing essential areas:

1. Take leadership in safety, health and wellbeing and demonstrate commitment
2. Identify and eliminate hazards at the workplace
3. Define targets and set goals to improve safety, health and wellbeing
4. Ensure a systematic approach to workplace safety and health
5. Use safe and healthy work equipment
6. Improve qualifications and competence through training and instruction
7. Invest in people, work together and motivate participation

I am sure that the growing application of even these basics will lead to a remarkable decrease of work accidents and work-related diseases in small businesses. This simple framework for success helps create a culture of prevention. If we can convince at least 1% of the small businesses owners not yet involved to organize prevention and apply at least basic measures, we will make the world of work a better one on a global scale.

Dr Sven Timm is a geologist and an environmental protection and industrial hygiene specialist. Timm is Vice President of the International Sections of the International Social Security Association (ISSA) on Information and Prevention Culture. His expertise is in co-operation in prevention on international OSH projects.

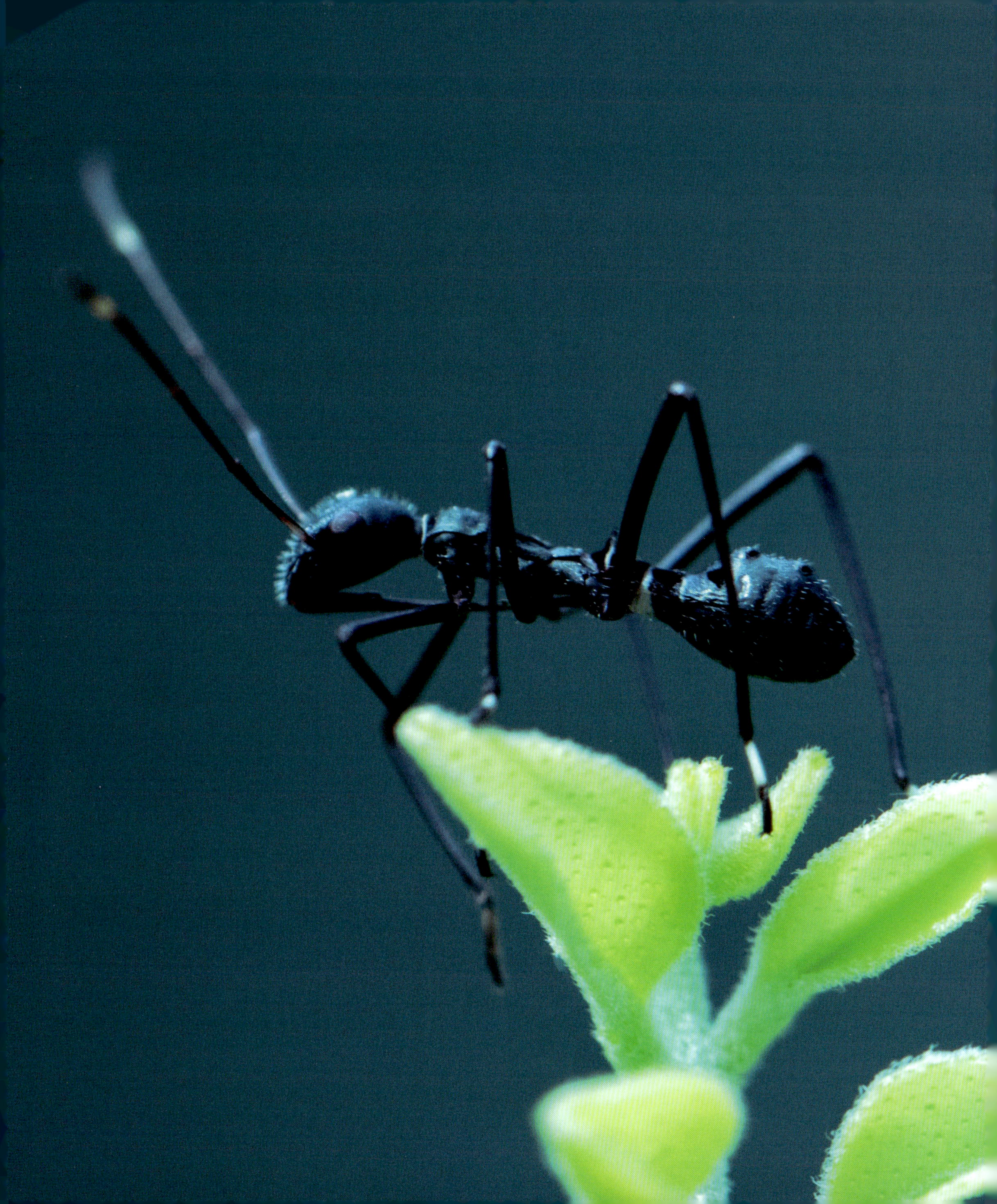

A lifetime goal of 'zero'

TWO SIDES OF THE SAME COIN

BERND TREICHEL

Today, the topic of safety, health and wellbeing is present on every TV channel, on social media, and at work, too. In 2020, we experienced a collective trauma, worrying about our health and our hygiene and caring for the health of our loved ones and friends – even more urgently if they are in the vulnerable population. As I write these lines, an estimated 90,000 healthcare workers are believed to have been infected with COVID-19 worldwide. It can be assumed that they contracted the virus at work, hence these cases should be considered as occupational diseases. In 2019, we might have been worried about reducing the number of occupational accidents.

In 2020, it is all about occupational health and hygiene – and what each of us can do to prevent ill-health. Is there a simple answer to this question?

Ever since its launch at the World Congress on Safety and Health at Work in 2017, ISSA has been promoting an approach that calls for a paradigm shift – a move away from rules, towards a participatory approach, towards leadership engagement in OSH and self-responsibility. 'Vision Zero' supports companies and organizations in developing a culture of prevention, in which safety and health lies at the core of all actions. The 7 Golden Rules that my colleague Sven Timm details on the previous page provide a roadmap for achieving Vision Zero. These Golden Rules are embedded in a self-assessment guide by ISSA for managers to identify areas for improvement within their enterprises. They have been field-tested and were found not only to be successful but also simple enough for managers to be able to confidently communicate them to their personnel.

Business leaders have introduced real change in their companies with the help of the new thinking of "zero". The Malaysian Social Security Organization runs a national Vision Zero strategy; a company in Pakistan displays big banners of the 7 Golden Rules in each of their production sites; and many employers all over the world have ceased the opportunity to discuss the 7 Golden Rules with their staff to boost employee participation, productivity and health and safety on site.

Vision Zero can improve the health and safety of companies large or small. Some may appear to be common sense – using safe equipment and identifying hazards, for example – yet research indicates that accountability via a checklist like the 7 Golden Rules makes 'common sense' work.

The saying "the way we do things around here" is a short definition of a company culture. However, because of COVID-19, "the way we do things around here" has changed significantly. We can transition into this new culture with the help of the Vision Zero values and its participatory approach among employers and workers at the workplace. While the concept of Vision Zero may appear opposing to the concept of One Percent Safer, they are actually two sides of the same coin. Leaders must recognize that by making small choices every day – to follow an actionable list of rules, to adapt to a new cultural focus on hygiene and wellbeing – we work toward a lifetime goal of 'zero.'

Every challenge we face in the world of work shows us how important it is to work together towards the same goals of zero accidents at work and zero occupational diseases.

Bernd Treichel joined the International Labour Office as a Labour Inspection Expert before moving to the Council of the Baltic Sea States and then the International Social Security Association (ISSA), as Senior Prevention Expert. He's a member of the International Organizing Committee of the World Congress on Safety and Health at Work and manages the Vision Zero Strategy.

BEING THERE (OR NOT)

NICK TURNER

This essay is about remembering where ideas come from, using the example of how being there (or not) gives us insight into understanding safety at work. This requires reconnecting the importance of employee presence (being there) and absence (not being there) through recalling findings from old – and often forgotten research – on safety and relinking it to contemporary knowledge.

Overlooking the demure beginnings of ideas in an age of Google Scholar searches is understandable. However, in marshalling the best available psychological knowledge about employee safety in a more mindful way, one encounters early research about the predictors and consequences of work injuries.

Early research by John Hill and Eric Trist of the Tavistock Institute of Human Relations[1] investigates the link between employee injuries and employee absence and illuminates ideas that modern safety research capitalizes on. Published in the journal *Human Relations,* this program of research has largely been forgotten in organizational psychology and safety research.

I'll identify four ways in which the Tavistock Institute research foreshadowed key organizational psychological and safety-related issues: two ways by 'being there' and two ways by 'not being there'.

First, in terms of 'being there', while Hill and Trist (Ibid.) emphasized individual differences in injury causation (accident proneness with psychoanalytic roots), they also emphasized the importance for safety of the relationship between employees and their employer, and the consequences of the misfit between the two. These ideas predate what became vast research literatures on organizational commitment and person-organization fit. Trist and his American collaborators[2] including Paul Goodman[2] later used the importance of 'being there' to explore the safety benefits of autonomous teams (vs. less autonomous teams) in coal mines, and how permanent employees 'being there' (not absent) generated job knowledge and familiarity that kept miners safe[3]. Second, Hill and Trist offered early organizational cultural explanations for injury occurrence, which may come as a surprise to those who see the idea of safety climate originating almost thirty years later with Dov Zohar[4].

Third, in terms of 'not being there', despite Hill and Trist creating the socio-technical approach with their research, it paid remarkably little attention to the technical side of injury causation. This seems strange, given they were studying steel workers who faced *significant* physical danger, regardless of how their work was organized. Fourth, the Hill and Trist research discounted injury severity, treating injuries like homogenous phenomena. Hill and Trist attributed various motivations for employee absence from injury, but largely forgot how the physical nature of the injury would likely predict employee absence frequency and duration.

Fast forward over half a century, contemporary meta-analyses[5] that synthesize safety research conclude that employee-organization relationships, safety climate, and interpersonal interactions with supervisors and co-workers all play important roles, if not the most important roles, as psychological predictors of work injuries.

Looking back can help to contextualize not only our present understanding of work safety research, but also what questions we might ask, and seek to answer, in the future.

1. Hill, J.M.M., & Trist, E.L. (1953). A consideration of industrial accidents as a means of withdrawal from the work situation. A study of their relation to other absences in an iron and steel works. *Human Relations, 6,* 357-380.
Hill, J.M.M., & Trist, E.L. (1955). Changes in accidents and other absences with length of service. *Human Relations, 8,* 121-136.
2. Trist, E. L., Susman, G. I., & Brown, G. R. (1977). An experiment in autonomous working in an American underground coal mine. *Human Relations, 30,* 201-236.
Trist, E.L., Higgin, G.W., Murray, H., & Pollock, A.B. (1963). *Organizational choice.* London, England: Tavistock Publications.

3. Goodman, P.S. (1979). *Assessing organizational change: The Rushton quality of work experiment.* New York: Wiley.
4. Goodman, P.S., & Garber, S. (1988). Absenteeism and accidents in a dangerous environment: Empirical analysis of underground coal mines. *Journal of Applied Psychology, 73,* 81-86.
5. Zohar, D. (1980). Safety climate in industrial organizations: Theoretical and applied implications. *Journal of Applied Psychology, 65,* 95-102.

Nick Turner is Professor of Organizational Behaviour and Distinguished Research Chair in Advanced Leadership at Haskayne School of Business, University of Calgary, Canada. He studies the predictors and consequences of 'healthy work', including leadership, work design, and occupational safety. He was editor-in-chief of the Tavistock Institute's *Human Relations* from 2017-2020.

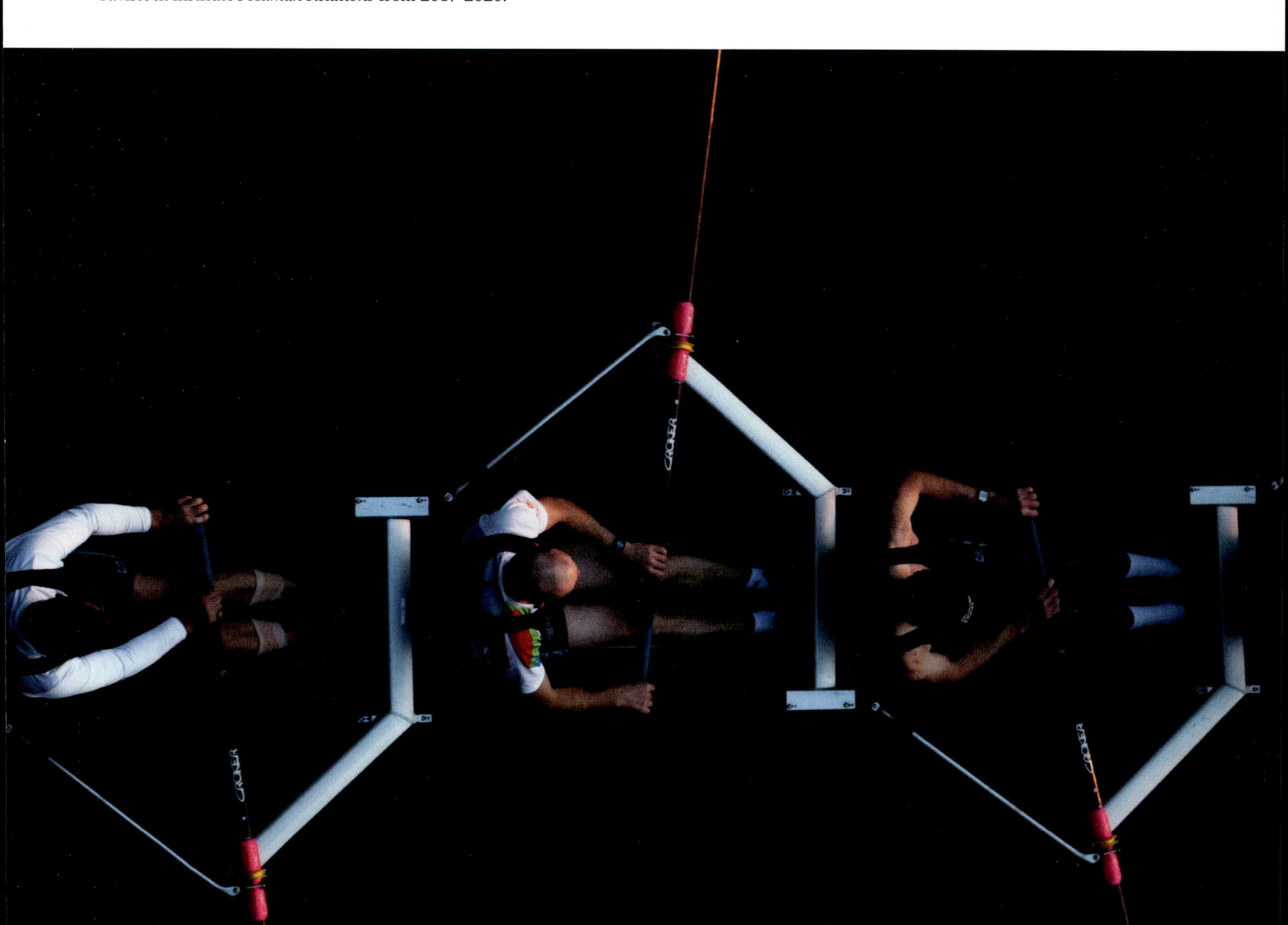

ABOLISHING THE ROLE OF SAFETY MANAGER

DAVIDE VASSALLO

The era of the safety manager is over. Certainly, their contribution in terms of lives saved, accidents averted and cost avoided is considerable, and indeed commendable. And yet, when considered in light of emerging trends in business, there is greater opportunity to improve safety outcomes by cultivating a workforce of *generalists* fluent in safety management and driving systemic organizational change.

Given the unprecedented scale, scope and speed of change within the business world, the focus on specialization in recent years has led to organizational rigidity, and dare I say, a sense of narrow-mindedness.

Generalists tend to draw inspiration from different disciplines, and to be more fluent in collaboration than specialists. They are more adept at viewing systems and processes holistically, as well as driving change throughout an organization. Even on a human level, the generalist can better identify and assuage the concerns of stakeholders throughout the company, building cohesion. These skills can help increase organizational flexibility, arming companies to face emerging challenges. As such, I believe that the era of the specialist is waning, while the era of the generalist is waxing.

While safety managers can be defined as specialists, I maintain that they are actually generalists; they cultivate effective working cultures, serve as advocates, engineer processes, initiate behaviour modification, influence others, and, most importantly, they lead organizations. Indeed, they engender many qualities that we seek in leaders and hold the mindset that would allow them to safeguard organizations as they evolve and grow. I say *leverage their talents* by promoting them into leadership positions.

Furthermore, safety managers are emblematic of a siloed approach to business. Beyond the typical bloat, expense and waste of time caused by silos, if we look at safety in particular, the relegation of accountability to one department or individual makes safety something to be enforced, not practiced habitually. Instead, safety should be led from the top, engineered into the operating system and felt throughout the entire organization. Siloing safety leads to plateauing, or even declining performance.

In the absence of safety managers, companies must instead focus on empowering employees to *own safety*. To achieve this, employees must be incentivized to work safely and abstain from cutting corners. This goes beyond compliance; education and skill-building are the best means of achieving systemic change.

Companies should also go beyond behaviour modification, focusing instead on risk awareness. By educating employees on the potential consequences of a given behaviour – positive or negative – it is possible to increase safety performance, and even drive operational discipline. Realising greater reliability and efficiency through operational discipline is the pretext for organizational agility.

While a career in safety management is certainly noble, I have personally seen the enormous potential of safety managers. Beyond their technical and operational prowess, they truly embody the humanity and goodwill of a company. I say delete this function, elevate these talented individuals, and leverage their capabilities to improve the performance, and even viability, of your company.

Davide Vassallo is CEO, DuPont Sustainable Solutions. He leads a global operations management consulting business to help organizations protect their employees and assets. Vassallo has published over 50 articles for newspapers, journals, magazines and conferences on risk management, leadership, carbon management and sustainable development.

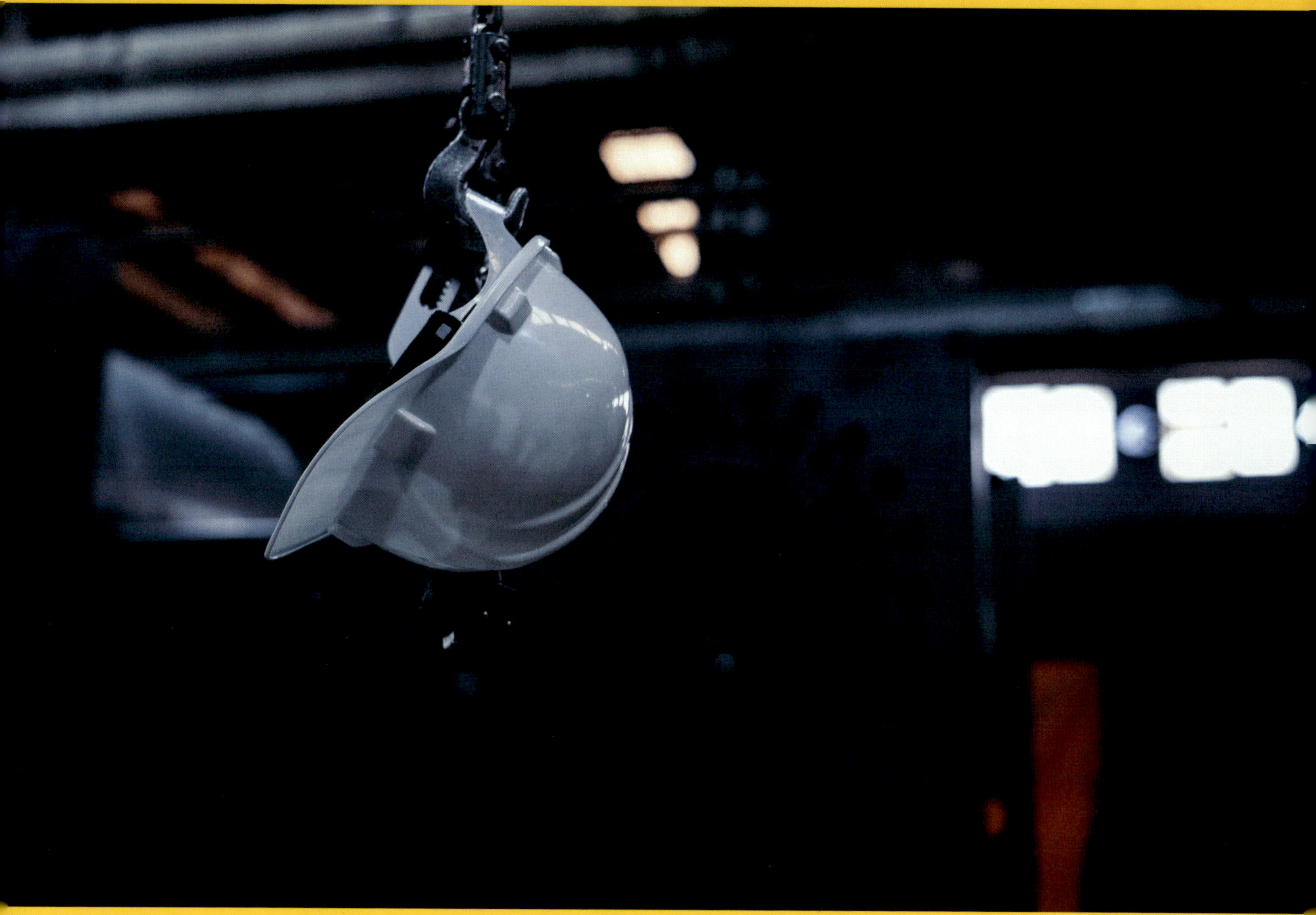

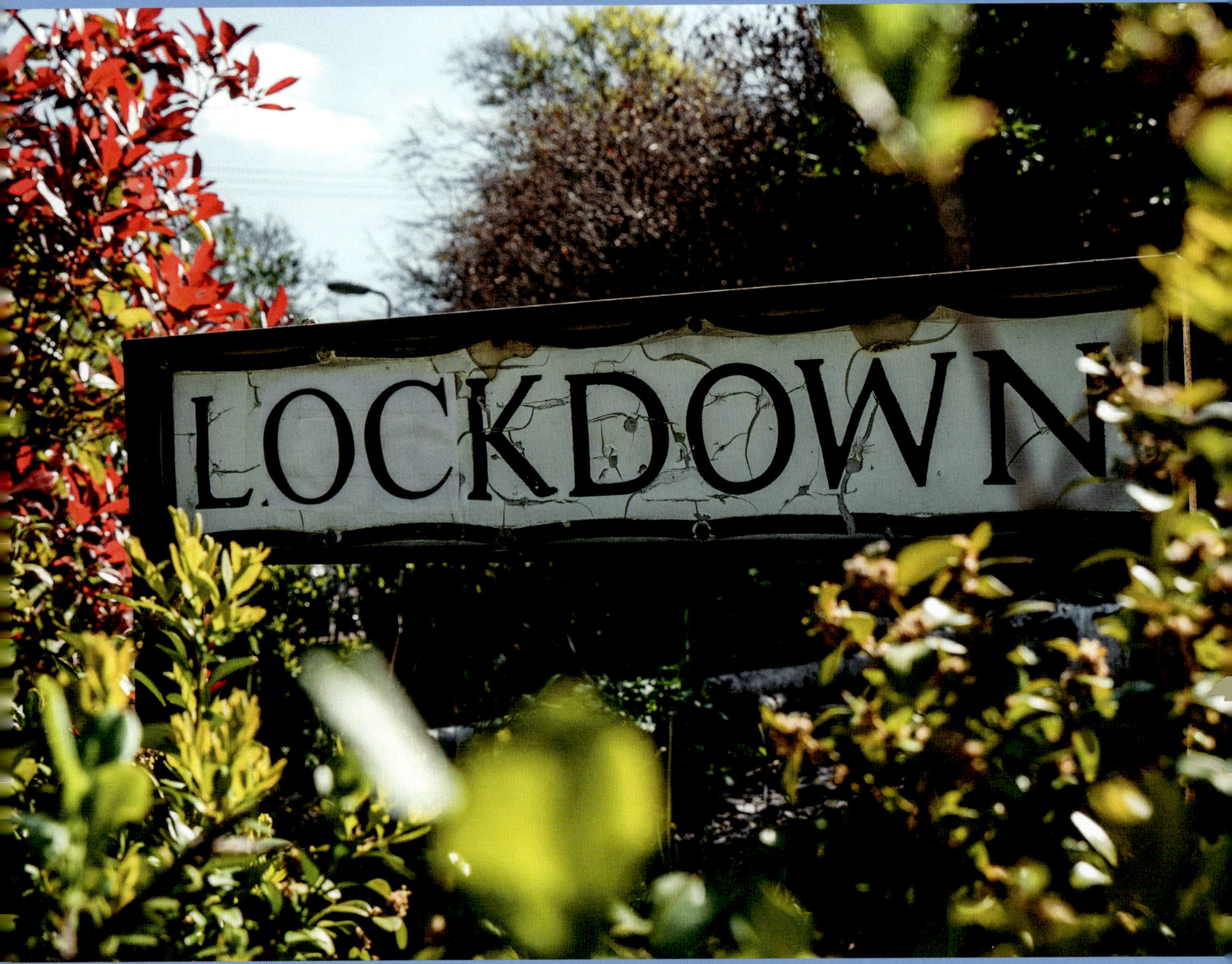
LOCKDOWN

GET RID OF THE MAN

BRUNO VERCKEN

At the time of writing, we are about to break the lockdown imposed by the COVID-19 pandemic. During the past nine weeks, health and safety professionals have not stopped working, in order to contain the spread of the coronavirus: masks, distancing, hand washing, work organization … We have done everything to keep people safe, and away from each other. We ended up sending them home, banning them – 60% of them – from their workplaces. Keep man away to protect man.

A recent exchange with my fellow directors of safety in major FMCG groups on road risks came to mind. For each of our companies, road mortality remains difficult to eradicate. Many years of work have made it possible to 'get under control' hundreds of factories (in terms of electrical risk, fire protection, working at height, chemicals, etc.), yet still the mortality of our salespeople or distribution teams on the road remains too high. The discussion focused on the progress of the driverless vehicle. It's a bright future within technical reach: no more drivers, to protect the drivers.

The risk prevention job is guided by the principle of keeping people away from danger. The ideal being to eradicate the danger, and failing to do so, to protect man and woman with equipment, procedures, training, PPE.

But we can also eradicate man from the equation. With COVID-19, we have all overwhelmingly chosen, and rightly for lack of vaccine, the latter.

In this phase of deconfinement and gradual exit from the pandemic, we are collectively at a crossroads. To build this 'new normal', the latest buzzword that holds all of our hopes for a more virtuous world, we have to choose between two options.

One is to continue our journey towards an 'all technical' workplace. It started long before the pandemic. But today there's a meteoric acceleration, because it's seen as a way to permanently remedy the problems of man in the workplace. Technology is an end in itself: 90% of 'new normal' projects proposed to me are tech solutions, like Google glasses, no-touch systems, robots, distance detectors, remote tests. Every day we wait for the miracle proposal from a start-up.

The other option is to choose man. The company needs the commitment of men and women, but their wellbeing depends on work, and there is no work without health and safety risks. In this regard, the pandemic has been an opportunity to pay attention, as never before, to anxiety and stress at work. Never has the coherent integration of safety, health and wellbeing at work seemed so natural and relevant.

So now, we are all collectively faced with this choice: to refuse the risk and bet everything on technology to pursue the Promethean dream of 'zero hazard', or to bet on man, with his fragility and his freedom.

We health and safety professionals can choose to become process safety technicians, or we can continue to protect humans and to protect work.

Bruno Vercken is Global Health, Safety and Working Conditions Director of DANONE. Until 2011, Vercken was Global HR Operations Director. He joined DANONE in 2001. He holds an MS engineering from Supelec and an MBA from INSEAD. He started his career with Airbus Industrie and spent 10 years in consulting.

THE QUIET REVOLUTION

LOUISE WARD

I want to start a revolution.

I'm a typical revolutionary, but the good news is that revolutions don't have to be about one person taking big actions. They can be achieved cumulatively with lots of people making small changes over a period of time.

I want to quietly revolutionize the approach to OSH. Systems and procedures all have their place, but I believe it's small actions that have the power to make a real difference.

There's a fantastic Ted Talk by Mark Bezos, a volunteer firefighter the US. He talks about arriving at a house fire to find that it's already under control, but he's asked to go and fetch some shoes for the owner, who has been evacuated in her nightclothes. This seems like such a small thing, not overtly heroic, but her letter later emphasizes the difference that this small act of kindness made to her.

On a larger scale, the BBC's "Blue Planet" program revolutionized thinking about plastic and its impact on the environment. That single program has driven many people to make small changes in their lives which will cumulatively make a massive, positive difference to our planet.

A revolution doesn't have to be about one person with a big idea. It can be a movement created by lots of people taking small actions on a regular basis, demonstrating through our actions that we care: about people, about the environment and about ourselves. These are our core values, those principles that guide our conduct, and that's where we need to focus to move the health and safety agenda forwards.

I work in health and safety because I believe everyone should go home safe, well and fit to enjoy life outside work. Each day presents opportunities to take small actions that can make this a reality, and it can be really small things. Don't walk by a spill on the floor, stop and wipe it up. It's a small thing that can make a big difference. It also sets a tone that encourages others to do the same. Reach out and chat to someone who seems withdrawn. That small act could provide support and also establishes a precedent. Make the effort to carry a reusable cup. These sound like small things, but if we all do them consistently, they build to become a huge movement.

This quiet approach to revelation has the potential to deliver a really sustainable change in our businesses, to become "just the way we do things around here", and I think this can help us to take workplace health, safety and sustainability to the next level.

So why not commit to doing at least one small thing every day that makes a difference? The lovely thing is that there's a personal benefit too, because when you do something positive it makes you feel good.

Let's make small change part of our everyday lives and build a sustainable culture of caring in our workplaces.

Louise Ward, BSc (Hons) CMIOSH HSEQ, is Director of Siemens Mobility. Ward has supported development of legislation, guidance and policy. She's a visiting lecturer at Middlesex University and supports degree apprenticeships. Ward writes for the trade press and co-authored a handbook on wellbeing published by Routledge.

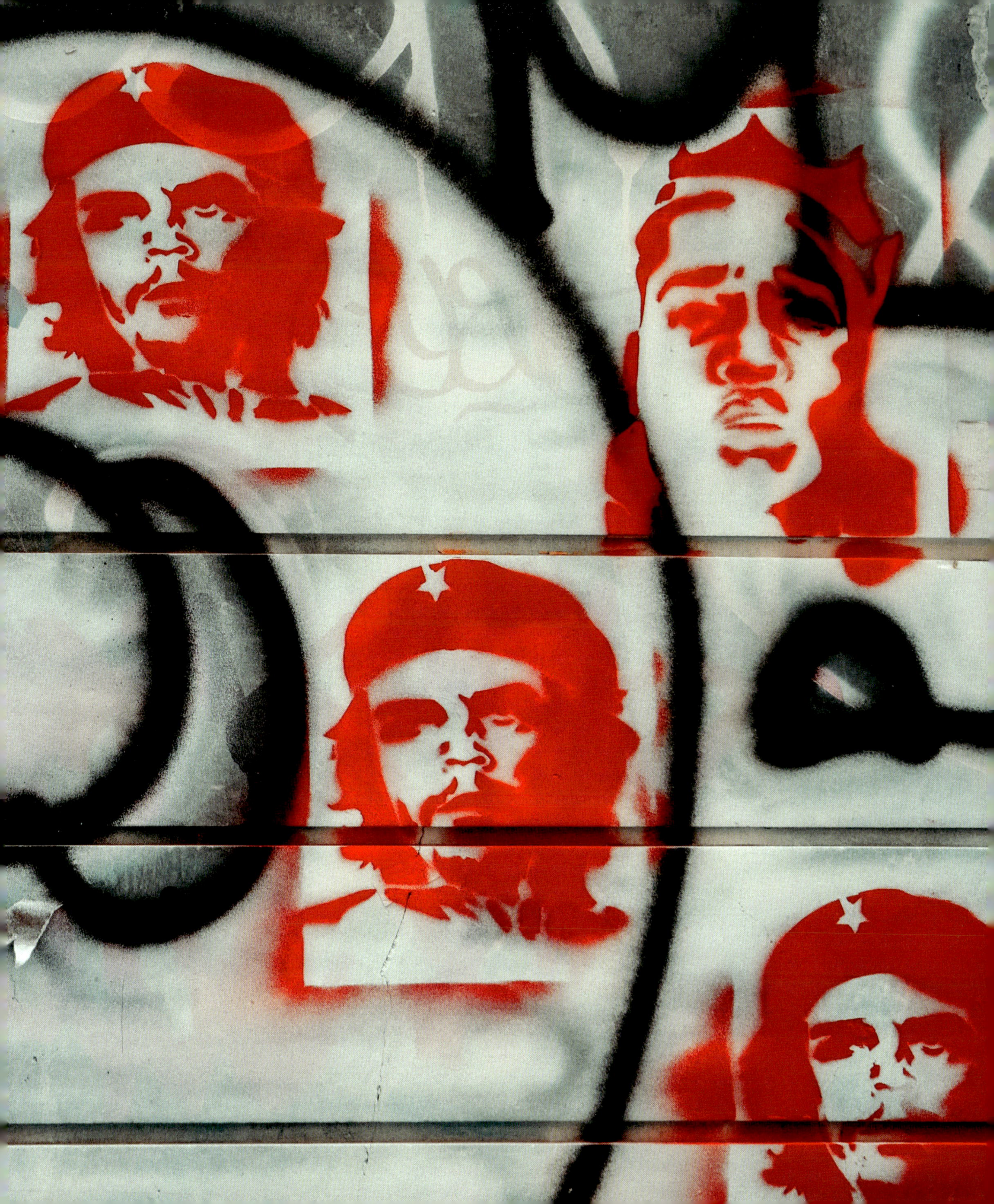

Think Once, Think Twice

THINK ONCE, THINK TWICE, THINK HEALTH

LAWRENCE WATERMAN

Some years ago, analysis showed that one of the reasons that motorcyclists on UK roads suffered so many serious, and often fatal, accidents was the way that car and truck drivers took no account of their presence and vulnerability. This stimulated a simple advertising campaign aimed at drivers, with the slogan:

"Think once, think twice, think bike."

That is the equivalent of what we need every worker to do, every supervisor, every manager – to create and maintain workplaces that don't damage people but instead enhance everyone's wellbeing – to think once, think twice, *think health.* Whenever a new shift starts, a new process is begun, new machinery used, a different work practice employed, a new team member joins, a new site starts operating ... As a matter of complete routine, we should ask, "What could this do to harm health? What can we do to make this healthier?" If answering that question requires some expert input – from hygienists, occupational health practitioners, industrial psychologists, then so be it.

This is not about making rules, issuing orders, drafting new procedures – it is a mindset that encourages everyone to think about their health, and how good work, well-designed and managed, can enhance their wellbeing. As we get better at preventing accidents, locally and globally, it is the opportunity to improve health that needs to be seized. Of course, the 2020 pandemic has stimulated interest in health matters, but the risks from toxic chemicals, noise and vibration, harmful dusts, and other health hazards cause more harm, more shortened and damaged lives than accidents at work. Being mindful of it and taking action will make more than one percent improvement. Workplaces and the wider environment will be so much more pleasant as a result.

Adopting this approach to health and wellbeing is also a great way to encourage engagement and leadership at every level of an organization from the desk, counter, ward and factory floor to the boardroom. Done well, it will greatly benefit the management of safety risks as well. This isn't a proposal for a top-down focus on health, rather, the aim should be the involvement of everyone in improving wellbeing.

Professor Lawrence Waterman OBE is Chairman of British Safety Council, Partner in Park Health and Safety, Visiting Professor at Loughborough University, recipient of the Institution of Civil Engineers medal for Safety and the RoSPA Distinguished Service Award. He was Head of H&S for construction for London 2012 Olympic and Paralympic Games.

SAFETY DOESN'T HAPPEN BY ACCIDENT

JOHN MARK WILLIAMS

We live in a world of risk. To improve safety, we need to manage that risk, and to manage risk, we need to recognize and understand it.

Agile organizations continually observe and assess. They embody awareness, transparency and insight – and gather intelligence for essential analytics.

They focus on iterative development – and evaluate at every stage. They reduce uncertainty to bite-sized risk – and resolve it safely at pace.

They know that greater agility means greater safety.

Be bold. Be involved.

Become an active part of a new and safer future.

State your commitment now to a 1% Safer world.

John Mark Williams is CEO of the Agile Business Consortium. He has been running businesses for more than 30 years across different countries and cultures. He's a believer in agile leadership and the responsibility of every manager to make the safety and well-being of their teams the foundation for fulfilment and success.

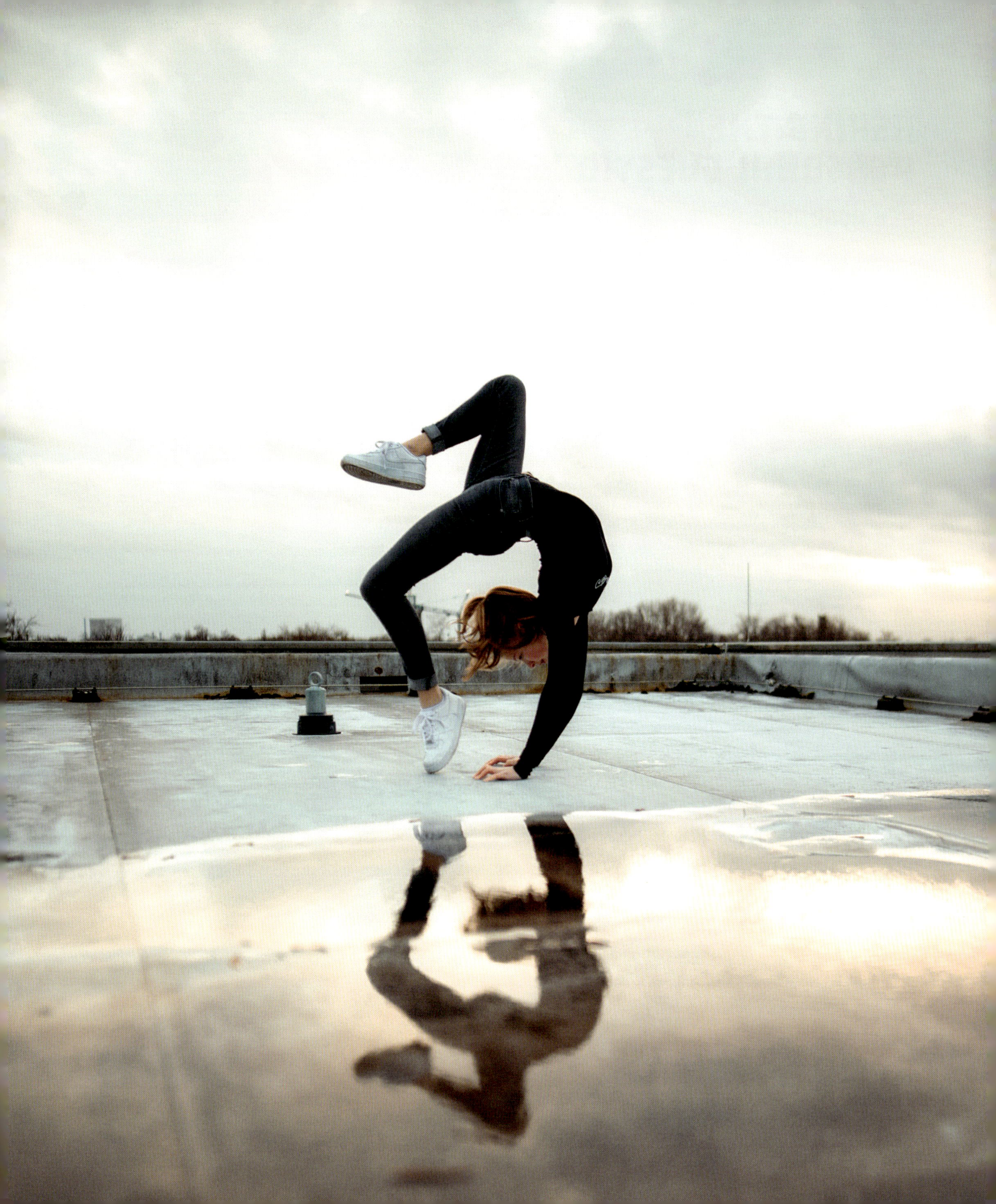

ASKING THE PERSONAL QUESTIONS

KEN WOODWARD

I have spent the last 25 years asking at least one million people in the workforces of hundreds of companies worldwide one major question, namely:

"Do you report all incidents/hazards/near misses/ minor injuries into the company reporting system?"

Usually management teams think their workforce would answer "yes" to this question, until they hear the answers I receive. In my experience worldwide, the answer is a resounding "no". It is clear that everybody does not always report, and this is a loss of what I consider 'gold dust', which is a precious material that will lead to both business and safety improvements. There are reasons why they do not always report, of course.

Here are just some of the necessary actions to take to put it right:

The leadership and management teams need to *personally* find out why their workforce does not report, and then set about building their own 'gold dust collection'. The Worldwide measurement is four reports per person per year.

Find out from your management system where there is a lack of reporting. Speak *personally* to a member of your workforce in that area and ask the following open questions:

"What difficulties are you experiencing in reporting incidents/hazards/near misses/minor injuries in your area?"

"How can you and your team overcome those difficulties?"

"What are the main challenges you face when working in this area?"

"What solutions do you have to put them right?"

Report these challenges and their solutions back to their line manager and the management team on site to encourage them to help and support their workforce to close each issue out for themselves.

Each leader and manager must later personally check whether the issue has been closed out and be held accountable for so doing as part of his or her tasks and targets. This will close the loop between managers and the workforce and encourage *personal communication* and continuous improvement.

And a final 'insight' for you: I can't wait for the leadership teams' remits for improving OHS to be created by the shop floor workforce in each company and presented to the Board for their support and buy-in, rather than the other way around. This single action will surely make us at least 'One Percent Safer!'

Ken Woodward OBE was blinded in a chemical explosion at work in 1990. His safety journey started then. Woodward has told his story to workforces in 90 countries and more than a million people. He is married with four children, seven grandchildren and step-grandchildren – none of whom he has ever seen.

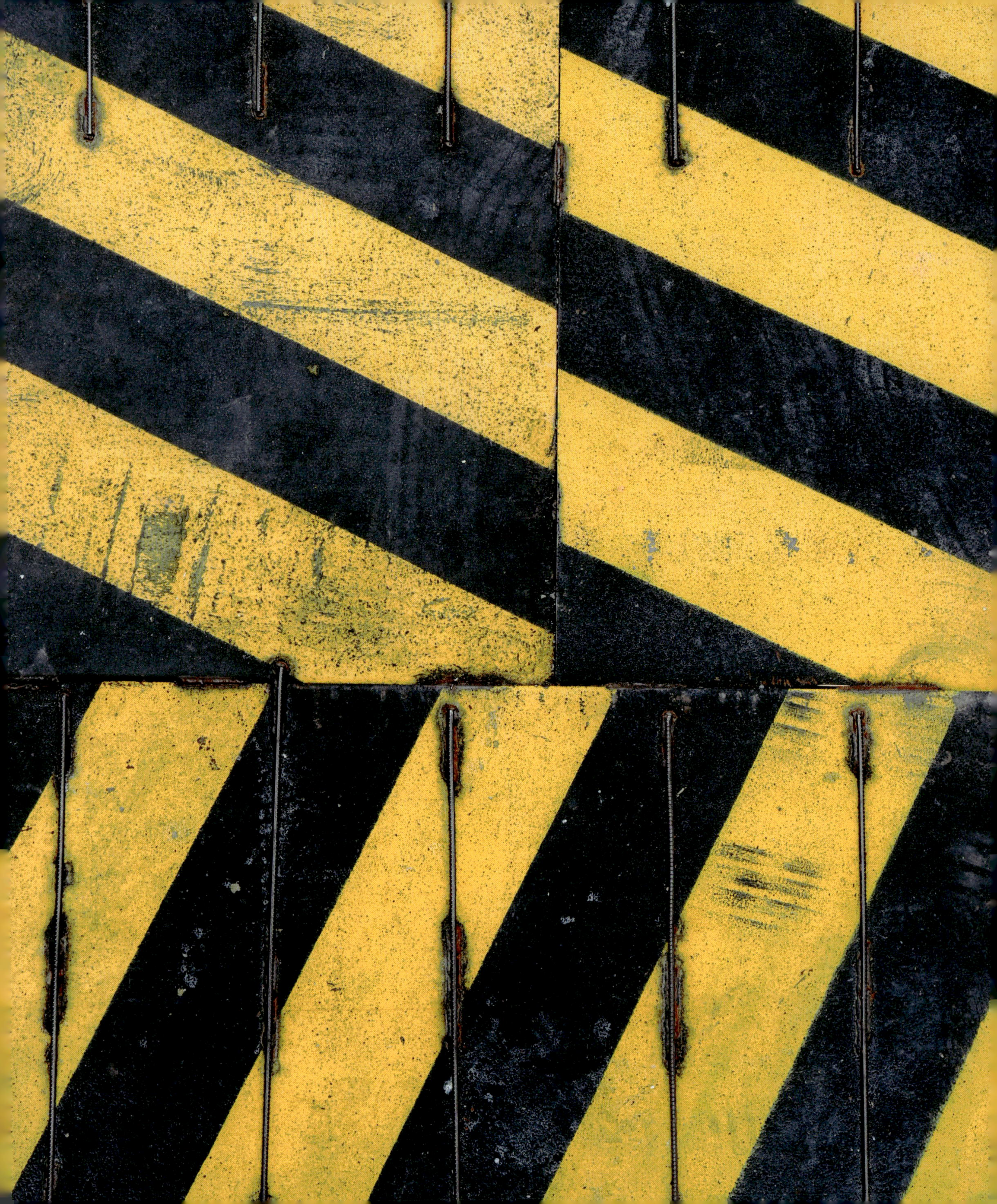

THE 'SURPRISE' OF SECONDARY HAZARDS

LOUIS WUSTEMANN

Over a couple of decades interviewing practitioners and regulators about what goes wrong in occupational safety and health, one theme I have noticed is how often OSH professionals are blindsided by hazards from work that are secondary to their main business activities.

That's not surprising; the risks associated with primary activities will always take priority. If you are employed to ensure the safety of workers at a chemicals plant, you are bound to focus your efforts on controlling exposure to hazardous substances. But the secondary hazards, the ones that involve only a handful of people, or that come with periodic maintenance or repair, involving work at height or in confined spaces perhaps, could be easy points of control failure, simply because they are not mainstream.

I once talked to an inspector who had investigated a food manufacturer, one of whose employees had lost the use of a hand after it was caught in a lathe at a canning plant. I told her I was surprised, because I knew the health and safety manager at the plant and rated him highly. The inspector said the production lines were some of the best controlled she had seen; the problem was that the company that supplied the manufacturing lines was no longer in business. So, to maintain a seven-day, three-shift operation with minimum downtime, the food maker had set up a little engineering workshop beside the factory floor to machine replacement parts for the canning lines. To reduce the health risk to workshop employees' skin from metalworking fluids, it had mandated wearing gloves at all times, which introduced the bigger risk of drawing-in injuries on the lathes.

"They knew how to run a food-making operation safely, but not a light-engineering one," said the inspector.

If there are any activities in your business, however small-scale, that create unusual hazard profiles or where you aren't completely confident you understand the best practice controls, it's worth making a little time to investigate. Find out what 'good' looks like, and check the risk assessments, method statements, equipment and training for any gaps or contradictions. Dedicating some time to those secondary hazards is another way to make your organization a little bit safer and better prepared for 'surprises'.

Louis Wustemann is a writer and editor specializing in health, safety and sustainability. He was formerly head of regulatory magazines at LexisNexis and was editor of *IOSH Magazine* and *Health and Safety at Work*.

USE PROACTIVE LEADING INDICATORS

GERARD ZWETSLOOT

To improve safety performance, good performance indicators can be of help. This is especially true when they give an insight into how improvements can be made, and when they are useful for benchmarking and mutual learning. Unfortunately, there are two main problems with performance indicators for safety:

1. Lagging indicators like the frequency of Lost Time Injuries ('LTI') are generally accepted and get most attention, but they do not predict the future, nor give you an idea of what should be done to improve performance. It therefore makes little sense to make people accountable for targets derived from lagging indicators (e.g. 10% reduction in LTIs, or zero LTIs), though it is common practice.
2. Leading indicators have – in principle – a predictive value, and it makes sense to make people accountable for realizing targets derived from leading indicators (e.g. increasing the percentage of technological innovations or organizational changes that are also meant to improve safety, see indicator 5.1 below).But there are only a few leading indicators based on good evidence, and there is not yet a harmonized set of leading indicators which hampers benchmarking and mutual learning.

To force a breakthrough, several companies that associated themselves with the Vision Zero strategy of the International Social Security Association (ISSA) suggested to develop a harmonized set of leading indicators, which should be evidence-based and useful for benchmarking across sectors and countries. I had the pleasure of leading a small team of key experts to develop a set of 14 proactive leading indicators. These are useful for any organization that aims for continual improvement, and applicable for safety, health and wellbeing at work. Companies from 20 countries shared their leading indicators and evidence with us; in the project we focused on those indicators that can trigger organizations to be more proactive. There are three options to measure each indicator (qualitative, semi-quantitative and quantitative). They can be used by organizations from all sectors and sizes.

14 proactive leading indicators for Safety, Health and Wellbeing at Work	
Visible leadership commitment	Competent leadership
Evaluating risk management	Learning from unplanned events
Workplace and job induction	Evaluating targeted programs
Pre-work briefings	Planning and organization of work
Innovation and change	Procurement
Initial training	Refresher training
Suggestions for improvement	Recognition and reward

(Source: VISION ZERO, *Proactive Leading Indicators. A guide to measure and manage safety, health and wellbeing at work*. 2020, Geneva: International Social Security Association, www.visionzero.global)

Dr Gerard I.J.M. Zwetsloot is an independent researcher/consultant based in the Netherlands and former honorary professor at universities in the Netherlands and the UK. His present focus is on Vision Zero, leadership and prevention culture. He received a lifetime achievement award from the Institute of Global Safety Promotion.

245

Well, I had to draw the line somewhere

AFTERWORD

Dear Reader,

In May 2020, during the pandemic, a close friend of my family was struck by the virus and died. It knocked the wind from my sails. Later that evening I received a note letting me know that in one of our client companies, a young father had been killed in a workplace accident. Moments later, a text from the CEO of another organization reported a similar message.

As we discussed in the early pages of this book, the numbers are huge: 2.78 million people killed by their work each year. Only when we start breaking that number down do they really start to resonate: 7,616 people dead every day, 317 every hour, one every ten seconds. Yet organizations rarely discuss safety in this way – it's usually only ever about *Accident Frequencies* and *Lost Time Injury Rates*. Real people suddenly fade out to become mere numbers.

"I am the one in ten, a number on a list.
Even though I don't exist.

Nobody knows me, even though I'm always there,
A statistical reminder
Of a world that doesn't care."

The words from the *UB40* song played over and over in my mind, as I recalled a moment at the end of 2019 where I stood in the palatial boardroom of one of the world's largest companies listening to the CEO tell me and his team that they had just had their 'best year yet' in terms of safety performance. I inquired how that was possible when they had suffered a fatal accident just a few months earlier. The room fell silent. The COO piped up *"But we don't include fatal accidents in our Lost Time Injury Rates"*. More awkward silence. I shared a slide with the face of the deceased worker and asked the execs to tell me who they were, where they lived. No-one knew. He was the one in ten, just a number on their list. As committed as most leaders are to worker safety, perhaps often the task of eliminating all injuries seems too vast, and as a result it can be hard to know where to start. So, on that cold, rainy Tuesday evening in May I hatched the idea of somehow harnessing the very best wisdom into some sort of compendium that any leader, anywhere, could access and put into practice in order to build some incremental progress towards a slightly safer, healthier world of work. Opening my 'little black book' and firing off a few emails with a rough outline of concept felt exciting at first, and then daunting to find, the next morning, no replies. I waited, a little nervously. The inbox remained quiet. At the end of the day a reply arrived, and then another, and another.

The contributions kept coming, from my 'professional heroes', eminent academics, top-notch scientists, business leaders, those with a story to share. Their cumulative power was both humbling and overwhelming at the same time. The mission (and subsequent title of this book), **one percent safer**, suggested to me that I needed to find 100 contributors. But I kept thinking of others that 'needed to be invited', so now, as you hold this in your hands, we've ended up with more than that. The rich variety of voices, perspectives, experiences and ideas feels both challenging and vital. As you've read the book you may have found yourself feeling a little discomfort – that's ok, just lean into it, believe in your own influence as a leader, stay focussed on the real prize: everyone going home without harm, every day.

There are no easy answers in these pages, but there certainly is a wealth of suggestions, perspectives and encouragement which may provoke some critical thinking or help generate some new ideas. Remember that *'doing less bad'* is not the same as *'doing more good'*. It really isn't about accident frequency rates, rules and zero tolerance. It's about a focus on the right inputs to get the right outputs. In a nutshell, it's about you *leading forward* and *creating safety*. Thank you for taking the first step and buying and reading this book. Now get out there and turn the ideas into action!

With love, trust, and hope,

Andrew

THE ONE PERCENT SAFER FOUNDATION

THIS BOOK WAS DREAMT UP, PRODUCED AND FINANCED INDEPENDENTLY FROM TRADITIONAL PUBLISHING MODELS.

All profits from the sale of the book go directly to the *One Percent Safer Foundation*, an independently-governed charitable fund created to make the world a safer place to work. The Foundation provides support to do this in two ways:

- The provision of practical help in the form of small grants or support with professional education and development to individual OSH practitioners who have lost their job as a result of the coronavirus pandemic

- The provision of small grants of 'seed capital' or support with education and development for third sector organizations, voluntary / charitable institutions, and not-for-profits (such as community organizations, social enterprises and co-operatives) who wish to begin a project in order to become **one percent safer**

We welcome applications to the *One Percent Safer Foundation* anytime.

To find out more please visit
www.onepercentsafer.com

DR ANDREW SHARMAN

Professor Doctor Andrew Sharman has been working in safety for nearly a quarter of a century, in more than 130 countries around the world, across all industry sectors – from the hallowed halls of the coolest Californian tech giants, to oil fields hidden in the frozen Siberian wilderness, from deep within platinum mines in Zimbabwe, to the mega-factories of China, with fashion hotshots and catwalk couture, to the top secret workshops of the world's fastest motorsport team, and from the brightest boardrooms of the world's biggest corporations, to the grittiest shop-floors. He's facilitated workshops for NGOs including the United Nations, the International Labour Organization, and the World Health Organization and spoken at over 500 professional summits, including the prestigious TED conferences.

He's picked up body parts and handed them to hopeful paramedics, testified under oath in courts of law, broken tragic news to loved ones, paid respects at worker's funerals, and discussed the lives of dead men and women with sorrowful Chief Executives. Over the years his vision has sometimes blurred by the horror of these moments, and his brain overloaded by death tolls and injury rates.

Hopelessness serves no purpose though: the only thing to do is to combine determination with practical wisdom earned over time.

Sharman is the author of 9 books on safety leadership, corporate culture, and wellbeing including the best-selling book on safety culture *From Accidents to Zero: a practical guide to improving your workplace safety culture* (find out more at www.FromAccidentsToZero.com).

He teaches at top-tier business schools around the world including *IMD* (Switzerland), *Caltech* (USA), and at the *European Centre for Executive Development* on the *INSEAD* campus in Fontainebleau, France, where he is Program Director for the *Leadership & Safety Culture* program. His work there is practical, not academic, as he uses a dynamic, provocative, hands-on approach to help senior leaders understand what safety really is, and how they can influence and drive a step-change in their corporate cultures. This style has been borne out of a lifelong fascination with how work actually gets done, how misunderstood safety really is, and a strong belief that just pushing harder brings about more problems than it actually solves. These interests also led to the idea of

ONE PERCENT SAFER.

Contact Andrew at **www.andrewsharman.com**

"Let a few things be repaired. A few is a lot. One thing repaired changes a thousand others."

John Berger, *Here is Where We Meet*